South University Library
Richmond Campus
2151 Old Brick Road
Glen Allen, Va 23060

MAR 0 6 2018

The Future of Disaster Management in the U.S.

"Disasters and emergency management reflect the broad policy trend toward centralization and the demise of federalism in the U.S. This volume seeks to reverse this course and offers a timely and much-needed vision of a less centrally planned and financed, and hopefully more robust and effective, approach to emergency management."
– *Dan S. Sutter, Troy University, U.S.A.*

U.S. congressional debates over the last few years have highlighted a paradox: although research demonstrates that emergencies are most effectively managed at the local level, fiscal support and programmatic management in response to disasters has shifted to the federal level. While the growing complexity of catastrophes may overwhelm local capacities and would seem to necessitate more federal engagement, can a federal approach be sustainable, and can it contribute to local capacity-building?

This timely book examines local capacity-building as well as the current legal, policy, and fiscal framework for disaster management, questioning some of the fundamentals of the current system, exploring whether accountability and responsibilities are correctly placed, offering alternative models, and taking stock of the current practices that reflect an effective use of resources in a complex emergency management system. *The Future of Disaster Management in the U.S.* will be of interest to disaster and emergency managers as well as public servants and policy-makers at all levels tasked with responding to increasingly complex catastrophes of all kinds.

Amy LePore is President of Anthem Planning, LLC and past President of the Maryland Emergency Management Association. In 2015 she earned a Ph.D. in Urban Affairs and Public Policy from the University of Delaware, U.S.A. Her research specialization is in federalism, dependence, and emergency management.

American Society for Public Administration
Series in Public Administration & Public Policy
David H. Rosenbloom, Ph.D.
Editor-in-Chief

Mission: Throughout its history, ASPA has sought to be true to its founding principles of promoting scholarship and professionalism within the public service. The ASPA Book Series on Public Administration and Public Policy publishes books that increase national and international interest for public administration and which discuss practical or cutting edge topics in engaging ways of interest to practitioners, policy makers, and those concerned with bringing scholarship to the practice of public administration.

For a full list of titles in this series, please visit
www.routledge.com/series/AUEASPSERPUB

The Future of Disaster Management in the U.S.
Rethinking Legislation, Policy, and Finance
Edited by Amy LePore

Adaptive Administration
Practice Strategies for Dealing with Constant Change in Public Administration and Policy
Ferd H. Mitchell and Cheryl C. Mitchell

Non-Profit Organizations
Real Issues for Public Administrators
Nicolas A. Valcik, Teodoro J. Benavides, and Kimberly Scruton

Sustaining the States
The Fiscal Viability of American State Governments
Marilyn Marks Rubin and Katherine G. Willoughby

Using the "Narcotrafico" Threat to Build Public Administration Capacity between the US and Mexico
Donald E. Klingner and Roberto Moreno Espinosa

Environmental Policymaking and Stakeholder Collaboration
Theory and Practice
Shannon K. Orr

The Future of Disaster Management in the U.S.
Rethinking Legislation, Policy, and Finance

Edited by Amy LePore

Routledge
Taylor & Francis Group
NEW YORK AND LONDON

First published 2017
by Routledge
711 Third Avenue, New York, NY 10017

and by Routledge
2 Park Square, Milton Park, Abingdon, Oxon OX14 4RN

Routledge is an imprint of the Taylor & Francis Group, an informa business

© 2017 Taylor & Francis

The right of the editor to be identified as the author of the editorial material, and of the authors for their individual chapters, has been asserted in accordance with sections 77 and 78 of the Copyright, Designs and Patents Act 1988.

All rights reserved. No part of this book may be reprinted or reproduced or utilised in any form or by any electronic, mechanical, or other means, now known or hereafter invented, including photocopying and recording, or in any information storage or retrieval system, without permission in writing from the publishers.

Trademark notice: Product or corporate names may be trademarks or registered trademarks, and are used only for identification and explanation without intent to infringe.

Library of Congress Cataloging in Publication Data
Names: LePore, Amy, editor.
Title: The future of disaster management in the U.S. : rethinking legislation, policy, and finance / edited by Amy LePore, Ph.D.
Other titles: Future of disaster management in the United States
Description: New York : Routledge is an imprint of the Taylor & Francis Group, an informa business, |2016| | Series: ASPA series in public administration & public policy | Includes index.
Identifiers: LCCN 2016029988 | ISBN 978-1-498-70001-6 (hbk : alk. paper) | ISBN 978-1-315-31077-0 (ebk)
Subjects: LCSH: Emergency management—Government policy—United States. | Emergency management—United States—Planning. | Emergency management—United States—Finance. | Disaster relief—Government policy—United States. | Disaster relief—United States—Planning. | Disaster relief—United States—Finance.
Classification: LCC HV551.3 .F87 2016 | DDC 363.34/80973—dc23
LC record available at https://lccn.loc.gov/2016029988

ISBN: (hbk) 978-1-498-70001-6
ISBN: (ebk) 978-1-315-31077-0

Typeset in Sabon
by FiSH Books Ltd, Enfield

For my family, with love.

Contents

List of Figures	ix
List of Tables	x
About the Contributors	xi
Acknowledgments	xv

1 The Centralization of Emergency Management   1
AMY LEPORE

SECTION I
Legislation   13

2 Intents and Outcomes for Local Businesses in Post-Disaster Contracting under the Stafford Act   15
CHRISTOPHER L. ATKINSON

3 Revising Federal Disaster Management Policy: Establishing an Officer in Charge   35
MARC LANDY AND JESSICA GOLEY

4 Assisting Individuals with Access and Functional Needs: The Intersection of Disabilities, Planning and Disaster Policy   60
MELISSA PINKE, STACEY MANN, AND ELIZABETH TODAK

SECTION II
Policy — 85

5 Local Recovery: How Robust Community Rebound Necessarily Comes from the Bottom Up — 87
EMILY CHAMLEE-WRIGHT, STEFANIE HAEFFELE-BALCH, AND VIRGIL HENRY STORR

6 Small Businesses as a Vulnerable Population — 102
MARK R. LANDAHL AND TONYA T. NEAVES

7 Managing Human Capital in Times of Crisis: The Role of Employees in Disaster Management — 127
STACEY C. MANN AND JONATHAN W. GADDY

SECTION III
Finance — 167

8 Major Disasters and Private Financing — 169
PETE VLOEDMAN

9 Financial Resiliency by Local Governments to Natural Disasters — 195
ROBERT BLAND, JESSECA E. SHORT, AND SIMON A. ANDREW

10 The Effects of Natural Disasters on Local-Government Finance — 209
ORKHAN ISMAYILOV AND SIMON A. ANDREW

Index — 230

Figures

2.1	Dates Contracts Signed, Hurricane Sandy-Related Contracting	25
8.1	Disasters Costing FEMA $500 Million or More	186
10.1	Sales Taxes Per Number of Business Establishments—Beaumont, TX	218
10.2	Sales Taxes Per Number of Business Establishments—Waco, TX	218
10.3	Sales Taxes Per Number of Business Establishments—Beaumont (city) and Jefferson (county), TX	219
10.4	Sales Taxes Per Number of Business Establishments—Waco (city) and McLennan (county), TX	219

Tables

2.1	Shifts in Paid Employees, Annual Payroll, and Total Establishments, 2012–13, New Jersey	22
2.2	Place of Contract Performance and Vendor State Comparison, Hurricane Sandy-Related Contracts	27
2.3	Extent Competed, Hurricane Sandy-Related Contracts	27
2.4	Most Common Vendors for Contract Awards, Hurricane Sandy-Related Contracts	28
6.1	Maryland Small Business Size Criteria	105
6.2	LSBR Program Size and Sales Criteria for Montgomery County, Maryland	105
6.3	Definitions of Small Business in Emergency Management Research	110
6.4	Factors Associated with Business Recovery	114
6.5	Factors Associated with Business Preparedness	115
6.6	Factors Associated with Business Mitigation Activities	116
7.1	Local Government Departments and Potential Emergency Planning Members	144
7.2	Frequency of Local Government Emergency Planning Meetings	146
8.1	Insured Losses Associated with Hurricanes by Saffir–Simpson, Category and Decade	170
8.2	Bonds for Each Area of Hurricane Concern	178
8.3	MultiCat Mexico 2009 Note Classes Description	178
8.4	MultiCat Mexico 2009-I Notes—Investor Distribution by Investor Type	179
8.5	Total Repetitive Flood Loss Properties in the NFIP: 1978–2011	182
8.6	History of Pre-Disaster Mitigation (PDM) Appropriations: FY 1997 to FY 2009	184
8.7	Requests, Appropriations and Supplemental Appropriations to the DRF: FY 2000–14	185
10.1	Demographic and Economic Characteristics: Beaumont and Waco	217

Contributors

Simon A. Andrew, Ph.D., is an Associate Professor in the Department of Public Administration at the University of North Texas. His research focuses on metropolitan governance and urban management—the role of governance institutions and human behavior/interactions in solving institutional collective action problems. He studies the challenges of developing and sustaining multi-stakeholder collaborations in the context of disaster planning and management. He is an expert in Social Network Analysis (SNA) and quantitative research methods.

Christopher L. Atkinson, Ph.D., is an independent researcher, with research interests in public management, policy studies, regulation, and emergency management. He has a Ph.D. in Public Administration from Florida Atlantic University and an MPA from George Washington University. He is the author of *Toward Resilient Communities: Examining the Impacts of Local Governments in Disasters*.

Robert Bland, Ph.D., is the Endowed Professor of Local Government and Faculty Director of the Center for Public Management at the University of North Texas. He is the author of several books, including *A Budgeting Guide for Local Government* (3rd edition). He was the recipient of the first Terrell Blodgett Academician Award and the Stephen B. Sweeney Academic Award from ICMA. In 2012, he was elected a fellow in the National Academy of Public Administration.

Emily Chamlee-Wright, Ph.D., is Provost and Dean of the College at Washington College. Chamlee-Wright is the editor of *Liberal Learning and the Art of Self Governance* (Routledge 2015) and author of *The Cultural and Political Economy of Recovery: Social Learning in a Post-Disaster Environment* (Routledge 2010).

Jonathan W. Gaddy is Director of the Calhoun County, Alabama Emergency Management Agency, where he has worked since 2007. He is currently participating in FEMA's National Emergency Management Executive Academy and is also a member of the National Disaster

Medical System (AL-1 DMAT). Mr. Gaddy is a master's candidate at the Naval Postgraduate School's Center for Homeland Defense and Security and completed undergraduate studies in geography and emergency management at Jacksonville State University.

Jessica Goley is a Ph.D. student of Political Theory and American Politics at Boston College. She graduated *magnum cum laude* with High Honors and a commendation in History from Kenyon College in May 2012. Among her many accomplishments, Jessica has been the recipient of the H.B. Earhart Fellowship, the T.W. Smith Fellowship and the Fortin and Bradley Summer Grants. During her time at Boston College, she has served as a model Teaching Assistant for Professor Marc Landy.

Stefanie Haeffele-Balch, Ph.D., is Research Fellow and Deputy Director of Academic Student Programs at the Mercatus Center at George Mason University. Haeffele-Balch is the coauthor of *Community Revival in the Wake of Disaster: Lessons in Local Entrepreneurship* (Palgrave 2015) along with Virgil Henry Storr and Laura E. Grube. She graduated in 2016 with a PhD in Economics from George Mason University.

Orkhan Ismayilov is a Ph.D. candidate at the department of Public Administration in the University of North Texas. His specialization is Financial Management. Mr. Ismayilov's area of interest is local government financial management, specifically economic recovery and development during and after natural disasters. Currently, Mr. Ismayilov is investigating the collaboration of local governments during natural disasters and the effect of natural disasters on local government revenue in the United States.

Mark R. Landahl, Ph.D., CEM® is a fifteen-year veteran of the Frederick County (MD) Sheriff's Office and adjunct faculty member at several universities. He earned his Ph.D. in Emergency Management from Oklahoma State University and has published in a number of academic journals including the *Journal of Homeland Security and Emergency Management* and *Homeland Security Affairs*. His current research focuses on survivor behavior in response to campus active-shooter events.

Marc Landy, Ph.D., is Professor of Political Science at Boston College. He has a Ph.D. in Government from Harvard University and writes on matters relating to the presidency, federalism and environmental and disaster-management policy. His disaster-management writings include: *Mega-Disasters and Federalism*, PAR: 2008 (supplement to Vol. 68); a review of presidential and congressional reports on Katrina (Publius: 38, 1, Winter 2008); and *Climate Adaptation and Federal Mega Disaster Policy: Lessons from Katrina, Issue Brief*, Resources for the Future Issue Brief (2010).

Amy LePore, Ph.D., is President of Anthem Planning, LLC and past President of the Maryland Emergency Management Association. In 2015 she earned a Ph.D. in Urban Affairs and Public Policy from the University of Delaware. Her research specialization is in federalism, dependence and emergency management.

Stacey C. Mann, Ph.D., is an Assistant Professor in the Department of Emergency Management at Jacksonville State University. She currently serves on the board of the American Society for Public Administration's Section on Emergency and Crisis Management as the journal coordinator. Her research interests include local government emergency planning, human capital management in disasters and crisis communication.

Tonya T. Neaves, Ph.D., is the Managing Director for the Centers on the Public Service with George Mason University's School of Policy, Government and International Affairs. She also serves as a faculty member in the Masters of Public Administration program and program director of its Emergency Management and Homeland Security graduate certificate. Dr. Neaves administers the Virginia Certified Public Manager® Program. Her research interests primarily focus on communal resiliency in the aftermath of disasters.

Melissa Pinke is in the final year of her doctoral degree in Emergency Management at Jacksonville State University. She has extensive experience in safety, health and emergency preparedness planning. One of her focus areas in developing emergency plans is to create awareness and simplify both actions and practices to include the needs of special populations so everyone can receive emergency information and can take protective actions quickly.

Jesseca E. Short, Ph.D., is a recent doctoral graduate from the Department of Public Administration at the University of North Texas. Within the field of public administration, her concentration is financial management. Her research interests include, but are not limited to, local government budgeting, financial resiliency, urban and metropolitan governance and disaster management.

Virgil Henry Storr, Ph.D., is Senior Research Fellow and Senior Director of Academic and Student Programs at the Mercatus Center at George Mason University, and Research Professor of Economics in the Department of Economics at George Mason University. Storr is the coauthor of *Community Revival in the Wake of Disaster: Lessons in Local Entrepreneurship* (Palgrave 2015), along with Stefanie Haeffele-Balch and Laura E. Grube, and author of *Understanding the Culture of Markets* (Routledge 2012).

Elizabeth Todak is a doctoral student in the Emergency Management program at Jacksonville State University, and holds a Master of Science in Emergency Management. She has worked in various positions in the public safety field for more than twenty years. Her research interests are varied, spanning healthcare coalition implementation, first-responder capabilities and the needs of practicing emergency managers.

Pete Vloedman is a pioneer in the Insurance-Linked Securities sector, beginning his work with the Chicago Board of Trade catastrophe futures product in 1991. He has more than twenty-five years of securitized and traditional reinsurance and asset management experience as a reinsurance company chairman, portfolio manager and underwriter and reinsurance broker. Pete holds an MBA from Columbia Business School as well as a BS in Marine Engineering from the U.S. Naval Academy.

Acknowledgments

The review process for the chapters was undertaken by a series of very talented individuals. I am very fortunate to have worked with them on this project.

Dr. Doug Goodman
Dylan R. Grifffith
Dr. William Hatcher
Paul V. LePore, Sr.
Dr. Patrick Roberts
Dr. Mark Skidmore
Dr. Daniel Sutter
Dr. Dong Keun Yoon

1 The Centralization of Emergency Management

Amy LePore

Centralization of Emergency Management

Whereas the performance of disaster-response mechanisms in the U.S. is often debated, rarely do we ever question the current delivery model. It is common to deliberate how successfully responding agencies performed, but we generally do not consider whether a federally centralized model of emergency management is most effective. What initially appears as a natural progression of institutional roles given lessons learned from Hurricane Andrew, 9/11 and Hurricanes Katrina and Sandy, has instead been a constant move forward of central planning since the 1990's. Along this path, there have been several notable contributing factors in emergency management which have bolstered a centralized model. These factors are mechanisms of political power and have been used to reward, realign and reengage entities involved in disaster response at all levels of government. Among them, and addressed in this chapter, are the increase in disaster declarations and the utility of the declaration itself as a political tool (Sobel and Garrett 2003; Sobel and Leeson 2006; Sylves and Buzas 2007). That the trend of centralization was exacerbated by the policy response to the attacks of September 11, 2001 is also noted in this chapter. FEMA's reduced capacity to act after natural disasters was seemingly a direct result of this transformation toward centralized planning for terrorism. In line with the theme of centralization, a direct dependence on federal grant funds to implement emergency management at the local level has been cultivated. While the advent of many of these grant programs has permitted the expansion of emergency management at the local level, it is difficult to know, given limited metrics, the actual effect of this arrangement of federal government as primary payer. These key points are matters for later in this chapter and will be offered in the context of the broader, deeper history of emergency-management centralization.

It is important to arrive at an answer to why this trend toward centralization is an issue worthy of examination and debate. Scholars generally agree that, especially after September 11, 2001, central planning of disaster response was a noted pattern (Roberts 2008; Scavo, Kearney and Kilroy 2008; Derthick 2009). FEMA itself has recognized this issue and its answer has been the Whole Community concept, as

> a means by which residents, emergency management practitioners, organizational and community leaders, and government officials can collectively understand and assess the needs of their respective communities and determine the best ways to organize and strengthen their assets, capacities, and interests.
> (Federal Emergency Management Agency 2011, p. 3)

However, this is not the only attempt by FEMA to attempt to harness better community participation and accountability. The Biggert–Waters Act of 2012 sought to reduce government subsidization of flood insurance rates, among other efforts, to shore up accounting. This was in response to several GAO reports listing the financial insolvency of the NFIP (Government Accountability Office 2005; Government Accountability Office 2011). Ultimately the Act was diluted by 2014 legislation which reduced the decentralizing effect. Still, there appears to be recognition by the federal government that a stronghold on responsibility for emergency management may not be the best path forward. Decidedly, if the pay arrangement does not change and responsibility continues to be centralized with the federal government, there is little chance that devolution of accountability toward lower levels of government will occur (Crabill and Rademacher 2012). What follows is an account of the two primary reasons that centralization and the proper roles of the levels of government should be closely examined for their impact on disaster management.

Alterations to the Intentions of Federalism

Dually focused on policy content and location of policy origination, critics of the current state of American federalism claim policy-making is increasingly centralized. Conversely, advocates of the broader federal role see increasing federal assistance and national policy-making as a method of solving widespread social problems and inequities. Each of these very different, very salient discussions regarding the role of government is bound not only to large and visible policy campaigns, but also to some of the most humbling and vulnerable of circumstances. Few events transpire which better frame the questions of federalism as clearly as disaster. Increasingly devastating and increasingly expensive, disasters serve to focus the nation's attention simultaneously on individual suffering and on the intricacies of government at work.

The level of centralization present in U.S. disaster management has breached the bounds of a traditional federalism. The disconnect between the intent of a federal system and the ways in which current policies are implemented at all levels is laid bare with the example of emergency management. Local governments—comprised of *we the people*, and to which all matters not addressed in the constitution should be left—are the recipients of billions of dollars of assistance to increase preparedness levels. Sullivan (2003, p. 1936) reminds us that the Tenth Amendment "endows the people with the right to choose and define their local government." In exercising this right, the people must also consider what priority programs should exist at the local level and practice the self-determination required to sustain these programs. However, in response to a 2013 survey of emergency managers nationwide, respondents overwhelmingly report that their office would not be sustained at the local level if federal dollars disappeared (Crabill 2015). Thus while there is no shortage of support from federal avenues, emergency management lacks the staying power of local support. This is the essential federalism disconnect that is further frayed by centralization. The federal mechanism for funding has aided in the removal of financial responsibility at the local level, and has dismantled local ability to make sound decisions (Crabill 2015). The states have varying degrees of involvement in programming funds, but have also seen their authority reduced by the Stafford Act, which places the federal government as primarily responsible for disaster. It is in this culture of dependence and misplaced roles that the future of emergency management rests.

Using Policy as Incentive

Federal grants have historically been used to achieve national goals. Sundquist and Davis (1969) argued that the relationships surrounding intergovernmental financial exchange have therefore been considered assistive in nature. Lovell (1981) points out, however, that federal grants-in-aid, often used to support local functions in lieu of local operating funds, have bred dependency. The debate over assistance versus dependence is alive and well. At the crux of this argument exists the question of whether or not federal policy—via grants—can effectively incentivize, and, if so, what is the societal tradeoff?

Federal grants provide leverage to the federal government for implementing national goals; the financing of local operations post-September 11, 2001 is an excellent example of this relationship. Most scholarship on this matter has focused primarily on centralized control of emergency management, and less on the impact of grants as incentive. That the primary disaster-response mechanisms still rest at the state and local levels is offered by Scavo, Kearney and Kilroy (2008) as proof that federal centralization has been resisted. However, this is contrary to the actual experience of local emergency management. Resistance to this effort of centralization would have likely been indicated by rejection of the grant funds that paid for this effort. Instead, state and local governments have taken in $35 billion in the decade after the 2002 inception of DHS grant programs (Federal Emergency Management Agency 2012). There is further evidence that not only grants, but funding in general, are expected from the federal government. As Birkland and DeYoung (2011) point out, one of the major complaints by the state and local governments regarding the federal response to the oil spill in the Gulf was that the "amount and speed of federal aid" was insufficient to meet their needs. Further, there is evidence that local governments alter their modes of operation based on the federal assistance. Donahue and Joyce (2001, p. 477) point out that "local governments still bear considerable responsibility for response and recovery efforts, but they may modify their activities in these areas to conform to federal criteria to secure as many resources as possible". This expectation of federal resources is not only occurring among practitioners but is further reflected in the literature, as Roberts (2008) believes that there was insufficient funding in the post-9/11 era. It remains to be seen if this process of modifying local activities to meet federal requirements for financial incentives is ultimately a best practice.

If local governments are not taking measures to increase their own resiliency in the form of ability to maintain the emergency-management function, can we soundly charge our nation's disaster resiliency to these offices? As federal policy has failed to incentivize strong and locally appropriate emergency management, Lovell (1981) might have argued instead that dependence has been bred. Further, if local entities do not believe that emergency management is a priority sufficient to provide funding, then we must consider: to what end is it maintained by taxpayers not proximal to the entity and instead rely on Federal taxes levied on jurisdictions from throughout the U.S.?

Overview of Methods of Centralization

History of Relief Spending

It is critical to show how state and local governments, and individual community members, have come to rely on the federal government for disaster-relief efforts. A brief history of how disaster relief has come to thrive is important to understanding current policy's inability to encourage independence from federal resources. Chiefly, this overview of the system of relief spending describes the way in which this relationship has been established, and how expectations have been cultivated, set and broken. An assessment of federal relief spending reveals an unexpected backdrop against which relief programming has grown. At first blush, one might assume that disaster relief is an extension of pre-existing social programs that assisted citizens who become homeless, displaced or unemployed. Instead, however, in the early years of the formation of the republic, relief spending to assist citizens plagued by disasters became the springboard for other types of general welfare spending (Dauber 2005). It is a wholly understandable association to have made, as certainly those affected by various social problems were pointed to as victims of circumstance in the same right as those victimized by hurricanes. But while debate over who was or was not actually a "victim" of disaster occurred with healthy vigor in the pre-New Deal era, there lacked a debate over the constitutionality of relief spending. Instead, disaster relief became the basis on which federal government spending and the welfare state was cultivated and grown. In *Sympathetic State*, Dauber (2005, p. 391) writes: "The very durability of disaster relief as a precedent had long made it a tempting ally for those seeking federal aid for other purposes..." Thus various occurrences of flood, fire and earthquake relief became the benchmarks for a way to frame problems such as illiteracy and poverty that were seen as imposing themselves on victims of circumstance. And while helpful at the time, in some sense, any opportunity to look at social afflictions in a different way was absconded by the early and little debated premise that most social problems can be solved via federal spending. Thus, the histories of federal disaster-relief funding and general welfare spending on social problems are intertwined. In each area we have seen the role of the federal government grow and, as in the case of disaster, overpromise and underdeliver. Because they are intertwined via their lengthy institutionalization, the histories of disaster relief and welfare spending inhibit interventions to limit and more accurately communicate the role and capability of the federal government. Therefore changes to the trend of centralization will not be simple.

This fundamental belief in general welfare spending on disaster, engrained in U.S. society from the earliest legislation, has ensured that so many disaster survivors had the opportunity to rebuild their lives. Unfortunately, this same programming has seen many citizens rebuild in the same hazard-prone areas with insufficient mitigation. On the surface, relief programming seems benign and benevolent, and in many cases, it is the only form of protection and recovery available to residents or businesses in the aftermath of disaster. The problem comes, however, when the message being sent is that the federal government will act as a safety net when, in fact, centralized government programming drives increased risk. An optimal outcome of disaster efforts at all levels would have been some transformation over the past 200 years toward risk reduction, community empowerment and state, local and individual responsibility. Instead, as an outgrowth of reliance, there has been little increase in independence from federal resources.

Event-Driven Centralization

After the terrorist attacks of September 11, 2001, an increased level of attention was paid to the management of domestic emergencies. It is often argued that planning for emergencies in the years between the attacks and Hurricane Katrina is best characterized as terror-focused, with reduced concentration on natural-hazards planning (Birkland and Waterman, 2008; Scavo, Kearney, and Kilroy, 2008). Several additional phenomena were at play between the attacks of 2001 and Hurricane Katrina. First, despite organizational upheaval in 2002, FEMA handled a series of four hurricanes in 2004 with very little public complaint, and thus there was little existing reason for concern about its capabilities (Derthick 2009). Second, despite years of warnings by meteorologists, government officials at all levels were caught off guard by this megastorm. In fact, similar storms had served as the basis for recent exercises and planning scenarios, yet there was a clear lack of awareness of the potential for destruction (Nathan and Landy 2009). Within this context, Hurricane Katrina devastated the Gulf Coast. Whereas there was little focus on the successes or failures of state and local government or individuals, the outcry about the role of the federal government was deafening. Birkland and DeYoung (2011) point out that this outcry was mirrored again after the Gulf Oil Spill. Despite the well-known shortcomings in FEMA's response to Katrina, Americans came to believe that FEMA has powers and capabilities that are far greater than those specified in the Stafford Act. That FEMA is the public face of all federal emergency response efforts caused the public—and many state and local officials—to believe that the federal government should issue some sort of "disaster declaration" for the oil spill. As has generally been the case, the failures at the federal level were resolved by writing new policy, which in turn only further solidified the centralized approach to emergency management in the U.S. (Crabill and Rademacher 2012). While there is now a turn toward all-hazards planning, the years during which funding was tied to terrorism-related preparations did indeed further centralize traditional emergency management (Birkland and Waterman 2008; Derthick 2009).

The Declaration Process

The years that followed the Stafford Act have seen a marked increase of declarations for disaster. While the declarations ranged in scope and nature, in 2012 the federal government spent $267,370,298 on public assistance—those funds reimbursed to state and local governments and non-profits—alone (Federal Emergency Management Agency, 2013). Surely, the new century has been marked by huge tragedies, including the attacks of September 11, 2001 and Hurricanes Katrina and Sandy, which account for much of the federal funding. Yet, this is but one method of funding transfers. While seemingly benevolent, the steady increase in declarations since the 1990's is reason enough to examine the decision-making process. Multiple studies have shown the distribution of federal funds post-disaster to assist with relief to be a highly political process (Sobel and Garrett 2003; Sylves and Buzas 2007). Further, Leeson and Sobel (2006) have identified a relationship between federal relief funding given to states and corruption. They write that this relationship is colored in part by the fact that "The incentive of political actors is to help themselves by distributing money in ways that benefit them and their political careers" (p. 8). Thus the literature points toward an unequal process which permits actors at the federal level to further centralize not only emergency-management efforts but also power in general terms.

Approaching Aspects of Centralization and Decentralization in this Volume

This book intends to provoke thought in terms of questioning some of the fundamentals of the current system and envisaging alternative models or exploring some aspects of the status quo. It seeks to forward perspectives on the current centralized model of preparing for or managing large-scale emergencies and examines whether accountability and responsibilities are correctly placed, as well as the extent to which current practices reflect an effective use of resources in a complex emergency-management system.

The broad categories of legislation, policy and finance capture critical elements concerning disaster management in the U.S. Further, these are the primary areas where the mechanics of centralization and decentralization can occur in a federal system. While many of the chapters included here touch on all three topics, there is a general theme in each which guided placement in the three sections. Among the chapters there are varying levels of technical information, lenses used for analysis and opinions on approaches to the editor's stance on centralization. Since there is varying opinion, this book presents ideas that tackle this centralization, further centralize, or offer a separate solution or change to the current mode of operation. Some are grand ideas, some propose minor changes and each has considerable merit. This book addresses a very disparate set of topics in emergency management. The authors' hard work on their chapters permits the reader to glean something from each piece but, more importantly, to consider all of the ideas together as alternatives to our current model.

Author Perspectives

Legislation

The section on legislation delves into the fundamental guiding documents which direct disaster-management efforts in the U.S. Primarily, the Stafford Act is addressed, given its role as authorizing legislation for so many disaster-management programs and practices. To address the matter of legislation fully, chapters in this section address a swathe of emerging issues in disaster management. Local business opportunity, bureaucratic command structures and functional needs planning are all discussed in this volume in order to illustrate the spectrum of legislation which dictates emergency-management efforts.

Chapters 2 and 3 address issues stemming from the current form of the Stafford Act. Christopher Atkinson's description of federal guidelines which offer preference to businesses in disaster-affected areas brings into question governments' ability to effectively legislate at such a granular level. The extent to which post-Sandy contracting data show an uneven distribution of business opportunity reinforces the problems associated with centralizing decisions of this nature. Far too removed from on-the-ground realities, the federal government cannot effectively select worthy business opportunities. Whether or not it should even be in such a position is another matter up for debate. Landy and Goley also address the issue of bureaucracy, and do so by suggesting change to the command structure in place after a mega-disaster. By the placing of an Officer in Charge, comparable to a general, the authors argue that local and state authority can be retained. The OIC would marshal required resources, cut red tape and be accountable for the federal role in recovery plans.

In Chapter 4, Stacey Mann, Melissa Pinke and Elizabeth Todak address legislation at all levels of government which dictates approaches to planning for special-needs populations. As a reminder of the expansive role of government in disaster, the charge of developing plans for each aspect of disaster-related disability needs is daunting. Compounding this complex undertaking is the expectation that government will hold an emergency assistance registry to use during disaster. Retaining this information appropriately from a privacy perspective, and ensuring that it is updated frequently, adds dimension to the expectations of government.

Policy

The section of this book which addresses policy-oriented questions is focused on individuals and communities most proximal to disaster. The authors have narrowed their research focus to those leaders, non-profits, local governments and businesses responsible for spearheading local response and recovery. In each of the three chapters critical points are considered and the outcomes of varying degrees of involvement across a spectrum of community and government are weighed. Matters of who should engage and when are woven within and among the chapters in order to solve the larger problem of drafting policy which calibrates authority appropriately. Indeed, each chapter offers key insights and a unique understanding of potential solutions along the range of centralization and decentralization.

In Chapter 5 Emily Chamlee-Wright, Stefanie Haeffele-Balch and Virgil Storr examine successful grassroots community engagement post-Katrina and how the conditions for this level of local leadership might be duplicated in future disasters. Relevant to this book is the broad question of whether or not the absence of federal government intervention in some aspects of community response will encourage further community engagement. Whereas the U.S. does not have substantial experience with devolution of authorities once held in a federal program, it is quite possible that leaders and organizations with a stake in local success will emerge. It is likely that federal government involvement has been so engrained in the culture of disaster response in the U.S. that it has just come to be an expectation. In fact, it is equally likely that this is the case across many policy areas. The authors' exploration of a changed format in which local individuals and organizations hold the decision-making authority, especially post-disaster, is a concrete step in the direction of culture change toward decentralization.

Local business, as one component of the community response mentioned above, plays a key role in response and recovery. In Chapter 6, Tonya Neaves and Mark Landahl have provided a thoughtful study on the ways in which small business has been defined in policy and research. In addition, the authors posit that the lens through which we see vulnerable populations should be applied as well to small business. Inclusion in the definition of "vulnerable population" should occur, they posit, based on the fact that existing government programs are already in place to provide additional assistance to businesses based on their size and the demographic makeup of the owners. In theory, this could advance the cause of small businesses in the pre- and post-disaster environments as it pertains to preparing for or recovering from an event. This is in part because, as the authors point out, tying financial incentive to mitigation or preparedness measures would have added benefit. The ability to recover quickly certainly assists in restoring the local economic base, but broader questions remain regarding a business' designation by the government as held to a set of characteristics based on ownership, geography or size.

Local-government response, the most proximal government response to disaster, is addressed by Stacey Mann and Jonathan Gaddy in Chapter 7. The authors examine human capital management implications and the related Human Resource Management (HRM) practices that can enable and empower efficient response. The significance of such a topic in this volume is twofold. First, it addresses the question of how a local government can properly engage post-disaster. Far past that of its emergency-management resources, local government has access to assets for response and recovery and is best situated to deliver services, given its knowledge of the local community. Second, the chapter addresses personnel matters that are internal to the local government and which relate to issues, such as labor standards, wages and time, which are also negotiated standards set by the federal government. As in the other two sections, it is not only emergency-management policy specifically which centralizes disaster-response efforts but also many, many peripheral polices instituted centrally by the federal government.

Finance

The final section of this volume focuses on finance policy as it pertains to emergency management. The three chapters span quite different topic areas: one is focused on federal policy and the other two focus on local-government financial practices. Each of the three chapters, however, aim at improving the ability to recover from disaster and building resiliency at all levels of government.

In Chapter 8, Pete Vloedman describes catastrophe bonds and their current use within the Mexican government to support disaster-management financial requirements. As the increase in the scope and cost of disaster in the U.S. has mirrored that of the rest of the international community, Vlodeman reminds us that it is imperative to look externally for best practices. Increasingly, the percentage of disaster borne by the U.S. federal government is growing, and shortfalls in the disaster-relief fund are becoming more evident. These deficits are a result of the randomness of disaster occurrences as well as the lack of sufficient forethought in budgeting for disaster relief. The author proposes the use of private risk financing as an addition to the current mechanisms of federal funding to alleviate existing budgetary concerns.

In Chapter 9, Robert Bland, Jesseca Short and Simon Andrew focus on the use of financial resiliency indicators for local governments and efforts to encourage sound community planning. Relevant to the theme of centralization is the authors' position that emergency-management efforts should be shared among governments as well as other sectors in order to support local operations locally—a turn away from federal-centric planning. Interestingly, the authors also encourage local government to focus on courting new businesses that have some level of resiliency built in; the primary example is construction, as this field has extensive utility in the post-disaster environment. Further, as a reminder of the risk often associated with living choices, Chapter 9 reminds local governments that the increase in population that bolsters the tax base may also become a liability if mitigation measures are not employed in flood-prone areas or areas prone to other types of disasters. With consideration to the fact that mitigation measures and the National Flood Insurance Program are federally funded, the focus here on local government resiliency as an alternative to centralization is well placed.

Finally, in Chapter 10, Orkhan Ismayilov and Simon Andrew approach the use of sales tax revenue as an area of concentration when discussing local-government financial resiliency. The authors write that while a primary focus of recovery should be to use available federal grants and loans, a government should also focus on promoting its workforce and examining existing sets of regulations in order to maintain an environment friendly for commerce. As an example of this the authors recommend rebranding the local-government work environment as a mechanism to retain personnel in the workforce as well as better preparatory budgeting for disaster.

Bibliography

Birkland, T. A., and S. E. DeYoung. 2011. "Emergency Response, Doctrinal Confusion, and Federalism in the Deepwater Horizon Oil Spill." *Publius: The Journal of Federalism* 41 (3): 471–493.

Birkland, T. A., and S. Waterman. 2008. "Is Federalism the Reason for Policy Failure in Hurricane Katrina?" *Publius: The Journal of Federalism* 38 (4): 692–714.

Brown, Orice Williams. 2011. "Flood Insurance—Public Policy Goals Provide a Framework for Reform: Testimony before the Committee on Banking, Housing, and Urban Affairs." www.gao.gov/new.items/d11670t.pdf (accessed August 23, 2016).

Crabill, Amy L. 2015. "The Effects of Federal Financial Assistance: Attitudes and Actions of Local Emergency Managers." Dissertation. ProQuest, May. http://udspace.udel.edu/bitstream/handle/19716/17054/2015_CrabillAmy_PhD.pdf;sequence=1 (accessed August 5, 2016).

Crabill, Amy L., and Yvonne Rademacher. 2012. "Breaking the Cycle of Reliance on Federal Help After Disasters." *Emergency Management Magazine.* www.emergencymgmt.com/disaster/Breaking-Reliance-Federal-Help-After-Disasters.html?page=2 (accessed August 5, 2016).

Dauber, Michele Landis. 2005. "The Sympathetic State." *Law and History Review* 23 (2): 387–442.

Derthick, Martha. 2009. "The Transformation that Fell Short: Bush, Federalism and Emergency Management." State University of New York, Nelson A. Rockefeller Institute of Government. www.rockinst.org/pdf/disaster_recovery/gulfgov/gulfgov_reports/2009-08-Transformation_That_Fell.pdf (accessed August 5, 2016).

Donahue, A. K., and P. G. Joyce. 2001. "A Framework for Analyzing Emergency

Management with an Application to Federal Budgeting." *Public Administration Review* 61 (6): 728–740.

Federal Emergency Management Agency. 2011. *A Whole Community Approach to Emergency Management: Principles, Themes, and Pathways for Action.* December. www.fema.gov/media-library-data/20130726-1813-25045-0649/whole_community_dec2011__2_.pdf (accessed August 5, 2016).

Federal Emergency Management Agency. 2012. *2012 The State of FEMA: Leading Forward: Go Big, Go Early, Go Fast, Be Smart.* www.fema.gov/pdf/about/state_of_fema/state_of_fema.pdf (accessed August 5, 2016).

Federal Emergency Management Agency. 2013. *Open Federal Data FEMA Public Assistance Program Detail.* Data.Gov. https://explore.data.gov/Federal-Government-Finances-and-Employment/FEMA-Public-Assistance-Funded-Projects-Detail/xkg4-9537 (accessed August 5, 2016).Jenkins, William O. 2005. "Challenges Facing the National Flood Insurance Program: Testimony before the Committee on Banking, Housing and Urban Affairs, U.S. Senate." www.gpo.gov/fdsys/pkg/GAOREPORTS-GAO-06-174T/pdf/GAOREPORTS-GAO-06-174T.pdf (accessed August 23, 2016).

Lovell, C. H. 1981. "Evolving Local Government Dependency." *Public Administration Review* 41: 189–202.

Nathan, R., and M. Landy. 2009. "Who's in Charge? Who Should Be? The Role of the Federal Government in Megadisasters: Based on Lessons from Hurricane Katrina." StateUniversity, Nelson A. Rockefeller Institute of Government. www.rockinst.org/pdf/disaster_recovery/gulfgov/gulfgov_reports/2009-06-02-Whos_in_Charge.PDF (accessed August 5, 2016)

Roberts, P. S. 2008. "Dispersed Federalism as a New Regional Governance for Homeland Security." *Publius: The Journal of Federalism* 38 (3): 416–443.

Scavo, C., R. C. Kearney, and R. J. Kilroy. 2008. "Challenges to Federalism: Homeland Security and Disaster Response." *Publius: The Journal of Federalism* 38 (1): 81–110.

Sobel, Russell S., and Peter T. Leeson. 2006. "Flirting with Disaster: The Inherent Problems with FEMA." *Policy Analysis* 573. http://object.cato.org/sites/cato.org/files/pubs/pdf/pa573.pdf (accessed August 5, 2016).

Sobel, Russell S., and Thomas A Garrett. 2003. "The Political Economy of FEMA Disaster Payments." *Economic Inquiry* 41 (3): 496–509.

Sullivan, J. 2003. "The Tenth Amendment and Local Government." *Yale Law Review* 112 (7): 1935–1942.

Sundquist, James L., and David W. Davis. 1969. *Making Federalism Work.* Washington D.C.: The Brookings Institution.

Sylves, Ricchard, and Zoltan I. Buzas. 2007. "Presidential Disaster Declaration Decisions, 1953–2003: What Influences Odds of Approval?" *State & Local Government Review* 39 (1): 3–15.

United States Governmental Accountability Office. "Report to Congressional Committees: High-Risk Series." February 2015. www.gao.gov/assets/670/668415.pdf (accessed August 5, 2016).

United States Governmental Accountability Office. "Report to the Ranking Member, Committee on Financial Services, House of Representatives; National Flood Insurance Program." March 2016. www.gao.gov/assets/680/675855.pdf (accessed August 5, 2016).

Section I
Legislation

2 Intents and Outcomes for Local Businesses in Post-Disaster Contracting under the Stafford Act

Christopher L. Atkinson

Introduction

The Robert T. Stafford Disaster Relief and Emergency Assistance Act (U.S.P.L. 100-707) of 1988 is the major law governing federally assisted disaster-related contracting in the U.S. The Act includes a stated preference in federally assisted post-disaster contracting expenditures "to the extent feasible and practicable... [for] those organizations, firms, and individuals residing or doing business primarily in the area affected by such major disaster or emergency" (42 USC §5150(a)(1)). The Act also allows for set-aside of contracts within a geographic area, and requires justification for contracts made outside the affected area. Concomitantly, in federal procurement the principle of full and open competition is a general directive, from executive agencies to grant-funded projects to state and local governments; this fundamental concept can work even in instances where other compelling government interest intervenes, as it does in the Act's preference language.

It would reasonably follow that such measures afford businesses in a disaster-affected area a means to contend for federally funded contracting opportunities in time of disaster and achieve success through the contract-award process. However, in practice, the process of bidding on and securing contracts can be complex, particularly for small businesses unfamiliar with bidding practices and requirements; further, procurement can be made using methods that fall short of the ideal of full and open competitive practices. Whereas this might be done for a variety of reasons, the net result may still be a limitation of opportunity for businesses most affected by hazard events—the same businesses that the Act's language seeks to engage.

This chapter explores the implications of the preference language, and procurement outcomes in light of a disaster case: 2012's Hurricane Sandy and its effects on New Jersey. The research question is: What do federal contracting data for Hurricane Sandy say about attention to local-business opportunity? Related to that point, what do the data say about transparency in contracting reports for disaster procurement? Analysis of these data show room for improvement for both local-level attention and transparency in reporting. It is recommended that additional guidance be provided to fully effectuate the spirit of the Act's provision: to support those businesses most in need of contracting opportunity as part of community and individual-level recovery. In particular, anything other than full and open competition in public spending related to response and recovery leads to predictably suboptimal outcomes from the perspective of local, small businesses. Familiarity and past experience figure heavily in contract decisions, for better or worse. Further, additional transparency is needed for contracting processes in disaster situations; inconsistent or non-existent data reporting must be addressed.

Policy and Literature Review

Full and Open Competition as a Central Principle of Public Procurement

The vision of the Federal Acquisition System has been "to deliver on a timely basis the best value product or service to the customer, while maintaining the public's trust and fulfilling public policy objectives" (FAR 1.102(a)). The system intends to "satisfy the customer" by

> Maximizing the use of commercial products and services ... Using contractors who have a track record of successful past performance or who demonstrate a current superior ability to perform ... Promoting competition ... [as well as] Minimiz[ing] administrative operating costs ... Conduct[ing] business with integrity, fairness, and openness, [and] Fulfill[ing] public policy objectives.
> (FAR 1.102(b))

One may detect, even in these initial pronouncements of intent, some points at cross-purposes with one another. Preferring contractors with track records might reduce the potential for new firms being brought into contracting. This could ultimately reduce capacity in a given field of expertise and leave government in the position of price-taker, having abrogated its responsibility to increase the pool of potential competitors. Minimizing operating costs might prevent unbundling of contracts, effectively preventing the broadening of contract opportunities available to smaller firms, since each contract requires administration. The idea of the public interest, and what that might mean in the context of public procurement, weighs heavily in notions of value and acceptability of costs.

Increasingly, governments are made to consider their relationships with the public; these interactions have at times been strained. There has been a fundamental distrust for what are believed to be wasteful practices or programs that fail to yield appreciable outcomes or provide good value for money. Implicit in the creation of an intricate structure like the Federal Acquisition System is a need for consistency in process and outcome. Attention must also be paid to fairness for accountability's sake, and recognition of the potential for, and prevention of, corrupt practices. The Federal Acquisition System acts through a compliance approach and means of standardization; it also encourages a 'business-like' orientation, in its proclivity to pleasing the client agency within certain regulatory bounds (Lawton, Rayner, and Lasthuizen 2013). Within acquisition regulations, the presence of limiting language might also serve as protection for contracting officers; if contracting officers follow the rules, it becomes more difficult to assert foul play in a protest review when a bidder loses and feels aggrieved.

The Federal Acquisition System also represents a primary point of interaction between business and government, and at times the differences between the sectors are very much in evidence. Although business and government have their flaws, they rely upon each other and feel negative effects when appropriate considerations are not made. "Corporations are intended to be profit-making institutions, and disregard of the ways in which businesses sustain social programs places the future of the society in peril" (Lehne 2006, p. 314). In times of crisis, the usual market mechanisms upon which the business world relies no longer seem to apply: communications can be difficult, if not impossible; supply chains become disrupted; business sites may become unavailable; and information-technology solutions may fail, leaving business to scramble for alternative approaches (Kildow 2011). In trying times, government can intervene in ways that make important positive differences for a community, be conspicuous through absence of meaningful involvement in disaster scenarios or provide for some level of service in between. Business must also recognize that government's involvement in society is purposeful, as it is government that maintains order and the rule of law—which is ultimately to the benefit of all concerned, including the business community.

Competition has been defined as "the struggle for commercial advantage; the effort or action of two or more commercial interests to obtain the same business from third parties" (Garner 2006, p. 122). Fair competition is distinguished in being "open, equitable, and just" (Ibid). In the interest of competition, FAR 6.101(a-b) notes requirements in "10 U.S.C. 2304 and 41 U.S.C. 3301" that,

> with certain limited exceptions ... contracting officers shall promote and provide for full and open competition in soliciting offers and awarding Government contracts ... through use of the competitive

procedure(s) contained in this subpart that are best suited to the circumstances of the contract action and consistent with the need to fulfill the Government's requirements efficiently.

FAR 6.200 et seq. allows for full and open competition after exclusion of sources. Keyes suggests this allowance exists in the FAR "because an agency is often 'locked in' on one source and it becomes very difficult for other sources to attempt to compete by incurring significant start up costs and necessary experience" (2000, pp. 103–104). Other exceptions, including national defense or reduced costs, are also allowed.

The federal government's requirement that all public procurement be subject to full and open competition unless otherwise allowed for by the applicable regulations is noteworthy. Reasons for taking another approach with procurement—for example, a non-competitive solicitation, commonly referred to as a sole source—must satisfy certain conditions. Further, the designation of sole source must be approved by increasingly responsible department leaders (FAR 6.304).

A White House advisory memorandum on improving public procurement notes, "Non-competitive contracts present a risk because there is not a direct market mechanism for setting the contract price" (Orszag 2009, p. 2). The memorandum goes on to reflect that

> Noncompetitive contracts enable agencies to address requirements that can only be satisfied by one source or that arise during emergencies when time allows for only limited consideration of offers ... agencies should make sure to limit use of these authorities to situations when they are truly appropriate.
>
> (Orszag 2009, p. 5)

The concepts of sole sourcing and unusual and compelling urgency deserve further attention because of their relevance to acquisition practices in disaster. The National Association of State Procurement Officials defines a sole-source procurement as "any contract entered into without a competitive process, based on a justification that only one known source exists or that only one single supplier can fulfill the requirements" (2015). Through the FAR, subpart 6.3, sole sourcing is provided under "other than full and open competition" authorities. Circumstances allowing for use of the authority are delineated in 6.302, and include only one responsible source and no other supplies or services will satisfy agency requirements; unusual and compelling urgency; specific expertise; national security; and public interest, among other reasons.

In terms of urgency, governments may make an assertion in disaster scenarios that "government would be seriously injured unless the agency is permitted to limit the number of responses from which it solicits bids or proposals" (Keyes 2000, p. 108). This would not be acceptable if the actual rationale was, for example, a lack of planning or a desire purely to keep costs low, because the absence of the market mechanism does not allow firms an opportunity to provide what could be a lower price for the same or similar product than the known sole vendor. For this reason, even in instances where urgency is an issue, agencies "are still required to request offers from as many potential sources as is practicable under the circumstances" (Keyes 2000, p. 108).

The public interest is contested space; the invocation of this value-laden term in regulatory text tending heavily toward the instrumentally rational suggests either a soothing of what may become a source of eventual public concern or a means of explaining away decisions on an as-necessary basis. Definitions of public interest show the disputed, context-based nature of the term and its use, as the following characterization suggests: "In a particular context, the public interest refers to the outcomes best-serving the long-run survival and well-being of a social collective construed as a 'public'" (Bozeman, in Lawton, Rayner, and Lasthuizen 2013, p. 35). The term lacks an objective basis; the values of those invoking the term come to the fore. For that reason, "on all matters governed by universal rights ... particularistic obligations hold sway," and this might suggest "cronyism, nepotism, and favouritism" (Etzioni, in Lawton, Rayner, and Lasthuizen 2013, p. 37). In a disaster situation, when social and community channels may become strained, employing provisions that reduce competition, even for good reason, might appear suspicious to communities. Having had so much taken through a disaster, survivors can feel hurt and angry, and may displace this emotion onto decisions made at society-wide levels. Affected populations may become alienated from their recovery. When decisions are not defensible, the anger may be justified, but regardless, transparency of decision-making is crucial. Contracting related to Hurricane Katrina evidenced such concerns (Dreazen and Opdyke 2005 and Gosier 2006 are examples).

Business Needs in Disasters and Government Responses through the Stafford Act

"Human settlements are susceptible to the harmful impacts of natural hazards," and this includes "lost jobs, business earnings, and tax revenues, as well as indirect losses caused by interruption of business and production" (Esnard and Sapat 2014, p. 65). Businesses may close, locations for operation may become unavailable and relocation options may be insufficient. Businesses often lack plans for how to address the impacts of crisis; they lack insurance and depend on government assistance, including SBA loans. Further, even if a business is able to reopen, business owners may find that their customer base has changed or, worse, no longer exists. Certain types of businesses may find themselves more vulnerable, and the socioeconomic vulnerability of owners may in turn impact the vulnerability of businesses, particularly small businesses (Esnard and Sapat 2014).

The Stafford Act provides that

> distribution of supplies, the processing of applications, and other relief and assistance activities shall be accomplished in an equitable and impartial manner, without discrimination on the grounds of race, color, religion, nationality, sex, age, disability, English proficiency, or economic status.
>
> (42 USC §5151(a))

The Act also requires

> transition [of] work performed under contracts in effect on the date on which the President declares the emergency or major disaster to organizations, firms, and individuals residing or doing business primarily in any area affected by the major disaster or emergency, unless the head of such agency determines that it is not feasible or practicable to do so.
>
> (42 USC §5150(b)(2))

Coupled with the stated preference for contracting with individuals and businesses in affected areas, under 42 USC §5150(a)(1), there is seemingly clear direction for public officials to make disaster-related work available to local, or even state or regional, businesses.

Subsequent to the failed government response in the case of Hurricane Katrina, adjustments to the disaster-contracting process were made, notably the establishment of the Disaster Acquisition Response Team (DART) in 2010. Whereas DART has been seen as a positive, supporting the "primary mission of deploying to provide disaster contracting support, such as contracting for blankets or debris removal," "FEMA does not have a process for prioritizing the team's work during disasters. Without such a process, FEMA is at risk of developing gaps in contract oversight during major disasters" (GAO 2015). Further, GAO's review, which centered on the period from 2005 to 2014, found some progress in implementing needed reforms, but did not find adequate attention to preference for local contractors, or transitioning existing contracts to local vendors. Some of the confusion hinged on misunderstanding about what terms, such as local, might mean in practice (GAO 2015).

Given that post-event work could be central to revitalization of an affected area, and act as a primary source of needed revenue to businesses adversely impacted by a hazard event, reviews of post-disaster contracting are worthwhile. The reality of post-disaster contracting might fall short of the ideal, whether or not for feasibility or practicability reasons. The Stafford Act's high-toned considerations of opportunity, so essential to the reduction of vulnerability and increase of an area's adaptive capacity, might be impacted severely, amounting more to soothing rhetoric than real prospects for contracting success.

Hurricane Sandy

Hurricane Sandy (October 22–29, 2012) formed in the Caribbean, moved through the Bahamas and grew substantially, before weakening somewhat and making landfall near Brigantine, New Jersey as a post-tropical storm (termed at the time a Superstorm, given its continued fierceness beyond tropical parameters). While winds weakened from 100 kts at Cuba, to 85 kts near the southeast U.S., to 70 kts at landfall, storm surge was ruinous. Sandy caused at least 147 deaths, including seventy-two in the northeastern U.S. (Blake *et al.* 2013). Other estimates differ: a U.S. Department of Housing and Urban Development report claimed that "at least 159 people in the United States were killed as either a direct or indirect result of Sandy" (HUD 2013). The storm adversely impacted 650,000 homes, more than 250,000 vehicles and 300,000 business properties (Rice and Dastagir 2013), and caused $71.4 billion in damage (Atlantic Oceanographic and Meteorological Laboratory 2014).

New Jersey was hit particularly hard. The statistics for destruction in the state are stunning:

> 346,000 housing units were damaged or destroyed … 100,000 new storm-related unemployment claims … 75% of New Jersey's small businesses were adversely affected … business losses are estimated at … $8.3 billion … at the peak, power outages left 2.4 million in the dark.
>
> (Smith 2013, pp. 1–2)

Damage in New Jersey alone was estimated at $29.4 billion (*Daily Mail* 2012).

Table 2.1 shows a comparison of County Business Patterns data from 2012 to 2013 (2016). Whereas the state as a whole increased its overall numbers in paid employees, annual payroll and total establishments, manufacturing was off considerably for each of those indicators. Real estate, professional services and management of companies and enterprises also saw declines in numbers of paid employees. As might be expected, construction, waste remediation and health care/social assistance all experienced big gains. However, given that at least some of these improvements were likely tied to disaster response/recovery, these are not

Table 2.1 Shifts in Paid Employees, Annual Payroll, and Total Establishments, 2012–13, New Jersey

Year NAICS code	NAICS code description	2012 Paid employees for pay period including March 12 (number)	2013 Paid employees for pay period including March 12 (number)	Change	2012 Annual payroll ($1,000)	2013 Annual payroll ($1,000)	Change	2012 Total establishments	2013 Total establishments	Change
—	Total for all sectors	3440470	3492216	51746	$189,910,527	$195,072,595	5,162,068	228289	230281	1992
11—	Agriculture, Forestry, Fishing and Hunting	h	2013	2013	$47,127	$44,574	−2,553	226	223	−3
21—	Mining, Quarrying, and Oil and Gas Extraction	1187	1329	142	$87,572	$93,136	5,564	77	77	0
22—	Utilities	18729	19059	330	$2,062,461	$2,063,233	772	412	387	−25
23—	Construction	130879	138817	7938	$8,124,353	$8,823,845	699,492	20115	20725	610
31—	Manufacturing	231143	221052	−10091	$14,579,930	$13,693,679	−886,251	7772	7587	−185
42—	Wholesale Trade	254465	257654	3189	$21,312,967	$22,324,291	1,011,324	14713	14555	−158
44—	Retail Trade	436542	445176	8634	$12,991,378	$13,038,250	46,872	31774	31711	−63
48—	Transportation and Warehousing	156707	158946	2239	$7,236,070	$7,746,426	510,356	7020	7171	151
51—	Information	90870	94715	3845	$8,449,329	$8,950,690	501,361	3603	3757	154
52—	Finance and Insurance	194502	198540	4038	$20,304,969	$20,860,950	555,981	12154	12177	23

Table 2.1 continued

53 —	Real Estate and Rental and Leasing	54730	53733	-997	$2,992,016	$2,919,316	-72,700	8693	8651	-42
54 —	Professional, Scientific, and Technical Services	311978	307495	-4483	$25,297,020	$26,327,275	1,030,255	29462	29651	189
55 —	Management of Companies and Enterprises	131949	130785	-1164	$15,933,929	$16,390,999	457,070	1394	1461	67
56 —	Administrative and Support and Waste Management and Remediation Services	281986	303691	21705	$10,483,165	$10,788,248	305,083	13630	13981	351
61 —	Educational Services	98988	100489	1501	$3,451,892	$3,605,267	153,375	3550	3665	115
62 —	Health Care and Social Assistance	546066	553578	7512	$25,047,442	$25,602,054	554,612	26924	27074	150
71 —	Arts, Entertainment, and Recreation	55560	56971	1411	$1,618,444	$1,619,013	569	3451	3476	25
72 —	Accommodation and Food Services	291694	297477	5783	$5,598,382	$5,755,503	157,121	20089	20462	373
81 —	Other Services (except Public Administration)	149640	150477	837	$4,288,402	$4,419,810	131,408	23079	23203	124
99 —	Industries not classified	175	219	44	$3,679	$6,036	2,357	151	287	136

Source: U.S. Census Bureau 2016

sustainable gains. Further, the last published year for County Business Patterns as of this writing is 2013, so it remains to be seen whether industry groups have stabilized and found some semblance of a new normal reflecting any permanent economic shifts that could have resulted from the storm. In terms of public, full-time state employees, not considering part-time staff, New Jersey saw a decline of 2,506 employees from 2012 to 2014 (132,767 in 2012 to 130,261 in 2014); full-time payroll increased from $791,775,480 in 2012 to $798,630,097 in 2014, as reported in the Census of Governments, State Government Employment and Payroll Data.[1]

Materials and Methods

The Federal Procurement Data System—Next Generation (FPDS-NG) collects information about and reports on expenditures associated with federal spending. Its report on Hurricane Sandy-related spending is the subject of this review. According to the FPDS-NG report for Sandy, as of January 14, 2016, $2,603,764,222 was associated with contract actions related to the storm. Of this amount, $627,599,489 went to small businesses (24.1%) (FPDS-NG 2016). The report includes this caveat:

> We believe [the report] represents a portion of the work that has been awarded to date. Many contracting offices supporting Hurricane Sandy particularly those relocated to the disaster recovery area, do not have access to their normal contract writing systems and thus have not been able to populate FPDS-NG contemporaneously with the contract awards they have made. Others have not had time to enter data due to the tempo of operations. It is impossible to estimate the impact this may have on the total numbers. As the operations tempo slows we expect that the data will be entered and thus the accuracy in terms of total contracts awarded, and dollars obligated, will increase.
> (FPDS-NG 2016)

As an additional point, the introduction page for FPDS-NG suggests that all potential sources of contracting information are not included; notably, purchase card data is left out, as is data from agencies not subject to FAR, interagency agreements, data associated with grants and a number of other sources of information.[2]

The research question is: What do federal contracting data for Hurricane Sandy say about attention to local-business opportunity? Related to that point, what do the data say about transparency in contracting reports for disaster procurement? This review analyzes Sandy-related contracting data. It is hypothesized that, aside from a concerning overuse of other than full and open competition in Sandy-related federal contracting, the outcomes do not suggest adequate attention to the Stafford Act's ideal of preference for doing business in the affected area. It is further posited that the level of utilization of other than full and open competition practices increases as project value increases; while there are many open contracting opportunities at lower contract values, large contracts are more likely to be reserved for certain vendors.

Results

The FPDS-NG report lists 5,105 contracting actions as of January 2016. The first contracts were signed on October 28, 2012; one contract was signed as recently as January 13, 2016, very close to the writing of this piece. In all, 50 per cent of contracting actions occurred on or before July 15, 2013. In the distribution of contracts signed, there are spikes near the end of fiscal years, which perhaps says more about bureaucratic processes than it does about responsiveness in the face of disaster. Noticeable increases occur around September of 2013 and 2014, with a smaller, but still apparent, increase noted in September 2015 (see Figure 2.1).

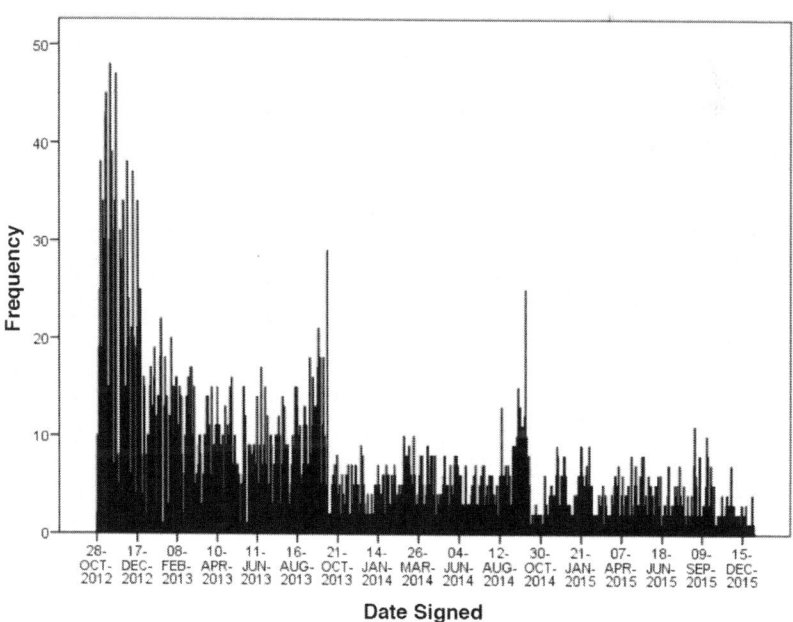

Figure 2.1 Dates Contracts Signed, Hurricane Sandy-Related Contracting
Source: Data from FPDS-NG 2016, figure created with SPSS

The Federal Emergency Management Agency is responsible for 1,524 of these contracts, or 29.9 per cent; the National Park Service had the next highest number of contracts, with 621 (12.2 per cent). The Coast Guard had 537 (10.5 per cent) and the Department of the Army had 534 (10.5 per cent); other federal agencies involved had less than 10 per cent each. Contract values ranged from no value noted to $128,169,947, excluding 659 listed as having a negative current contract value.

Whereas 24.1 per cent of contract value went to small businesses, according to the report, 55 per cent of contracts (2,806) were associated with small businesses. It stands to reason that most of the contracts received by small businesses were for smaller amounts, given business capacity. Contracts to "other than small businesses" (2,298) were fewer in number than small-business contracts, but they make the up the majority of contracting dollars (75.9 per cent).

Table 2.2 reflects place of contract performance compared with vendor state, for states experiencing direct effects from Hurricane Sandy (roughly the Mid-Atlantic from Virginia north through most of New England). As expected, New York and New Jersey showed the greatest number of contracts with those states as place of contract performance. Those states also did less well with regard to distribution of contracts to local firms, when not considering contract value. Given that each contract poses a contract opportunity, at least in theory, a balancing of contracts performed in affected states and commensurate distribution of contracts among firms in the affected area might be one estimation of achievement of the Stafford directive. However, taken as a group, one also sees considerable disparity in the expected high number of contracts performed in these states, and awarded vendor distribution to those states (a difference of nearly 18 per cent).

Total contract value is another consideration. In all, $182,421,309.82 in total contracting related to the storm was afforded New Jersey vendors, for contract value with base and all options.[3] This is 7 per cent of total dollars in the system as of this writing ($182,421,309.82/$2,603,764,222), lower than the percentage of all contracts for New Jersey as place of performance (23 per cent) and the percentages of contracts where New Jersey is the vendor state (14.3 per cent).

Table 2.3 shows the extent of competed-for contracts involved in the Hurricane Sandy response/recovery. Again, while full and open competition remains the ideal for public procurement, there was clearly much reason given to forego such approaches in practice in this case. A total of 422 contracts were not competed because other than full and open competition was "authorized by statute" (8.3 per cent) and 407 were not competed for reasons of urgency (8 per cent). Simplified Acquisition Procedure (SAP) streamlined competition was the reason given for 201 contracts (3.9 per cent), and only one source was the rationale in not

Table 2.2 Place of Contract Performance and Vendor State Comparison, Hurricane Sandy-Related Contracts

State	As Place of Contract Performance	% of all Contracts	As Vendor State	% of all Contracts	Comparison of contract performance location to vendor location (+ means vendor use % for the state outpaced % of contracts performed in that location)
Connecticut	128	2.5%	161	3.2%	0.7%
Delaware	41	0.8%	15	0.3%	−0.5%
District of Columbia	123	2.4%	152	3.0%	0.6%
Maryland	250	4.9%	272	5.3%	0.4%
Massachusetts	174	3.4%	316	6.2%	2.8%
New Jersey	1176	23.0%	731	14.3%	−8.7%
New York	1903	37.3%	897	17.6%	−19.7%
Pennsylvania	71	1.4%	167	3.3%	1.9%
Rhode Island	58	1.1%	31	0.6%	−0.5%
Vermont	1	0.0%	12	0.2%	0.2%
Virginia	304	6.0%	555	10.9%	4.9%
	4229	82.8%	3309	64.9%	−17.9%

Source: FPDS NG 2016

Table 2.3 Extent Competed, Hurricane Sandy-Related Contracts

	Frequency	Percent	Cumulative Percent
Competed Under SAP	1122	22	22
Competitive Delivery Order	3	0.1	22.1
Follow On To Competed Action	1	0	22.1
Full And Open Competition	1905	37.3	59.4
Full And Open Competition After Exclusion Of Sources	780	15.3	74.7
Not Available For Competition	363	7.1	81.8
Not Competed	577	11.3	93.1
Not Competed Under Sap	352	6.9	100
Total	5105	100	

Source: FPDS-NG 2016

competing 151 contracts (3 per cent). That said, many contract actions, at least according to these data, were informed by full and open competitive practices.

Table 2.4 shows the most commonly contracted vendors, event-wide.

At 10.5 per cent of the total number of Sandy-related contract actions, these ten vendors alone combine for $767,723,107 in total contracting value. This is about 29.5 per cent of total contracting dollars related to Hurricane Sandy, as reported by FPDS-NG. On the face of it, these results do not suggest parity. Among the top ten, New Jersey is represented only by Cellco Partnership (Verizon). It should be noted that "full and open competition" is frequently listed for contracts with large and common vendors. While the processes may well have met the letter of the law for full and open competition, process outcomes were predictable and favored established businesses.

As an example, AECOM received a contract for "emergency recovery and technical support services" from FEMA. On the AECOM website press release announcing the award, it is mentioned that the "contract is worth up to US$500 million during one base year and four option years.

Table 2.4 Most Common Vendors for Contract Awards, Hurricane Sandy-Related Contracts (Base and All Options)

Name of Vendor	Frequency	Percent	Total Value
Corporate Lodging Consultants Incorporated (Kansas)	125	2.4	21,531,905
Sunbelt Rentals, Inc. (South Carolina, New York, Washington (state), and Pennsylvania)	52	1	619,750
Calcedo Construction Corp. (New York)	51	1	42,763,484
Coastal Environmental Group Inc. (New York)	50	1	17,248,785
Cherokee Nation Technology Solutions, L.L.C. (Oklahoma)	46	0.9	3,759,793
Great Lakes Dredge and Dock Company, LLC (Illinois)	46	0.9	498,551,672
AECOM Recovery (Virginia)	46	0.9	141,901,816
IAP Worldwide Services, Inc. (Florida)	44	0.9	39,020,810
Cellco Partnership (incl. DBA Verizon) (New Jersey)	42	0.8	2,123,793
United Parcel Service Incorporated (Washington, DC)	38	0.7	201,300
Totals (Top 10)	540	10.5	767,723,107

Source: FPDS-NG 2016

This is AECOM's fifth consecutive Public Assistance Technical Assistance Contract from FEMA" and that "AECOM has a strong performance history with FEMA in its critical disaster-recovery missions across the United States and the U.S. territories" (quoting the AECOM chairman and CEO) (Dickard 2012). The press release observes the long-term partnership with FEMA. Great Lakes Dredge and Dock Company is the largest firm of its type in the U.S., with $731 million in revenue in 2013 (Hayes 2014). Again, this is a firm with well-established contracting ties with government. From the perspective of the procurement officer, it might be said that repeat customers and consistently high contract volume are a testament both to quality of work and to reliability—both important issues in response and recovery operations. As noted previously, the FAR has allowances for efficiency of procurement operations to fulfill government requirements, even with direction to fulfill public policy obligations of social import.

The point in raising this factor of procurement practice in the context of the Stafford Act is that the good intentions of legislation may sometimes be left behind in the reality of response and recovery. Federal acquisition rules allow for previous experience to weigh as a factor in award decisions, and there is a clear comfort level with certain firms over others in such awards. If the compelling interest in disaster contracting is to effectuate the good intentions of contracting locally, to encourage resilience and decrease vulnerability in areas directly affected, then such discretion on the part of procurement officers may need to be diminished in disaster scenarios. On the one hand, in taking such approaches, government seems to be behaving much like what would be expected from business—going with tried-and-true "partners" rather than exploring the market and possibly increasing capacity for competition's sake. As Tadelis (2012) suggested, the private sector is much more likely to use negotiated contracting built upon relationships than is the public sector, where open competitive solicitation is thought to reduce the potential for corruption while increasing transparency.

Discussion

Procurement outcomes for federal contracting in this case raise two different concerns as regards public contracting practices in times of disaster. For one, it is clear that the *urgency* claim is used frequently to justify purchases in times of disaster, but the decisions that are made under such cover might cut into the public interest of local contracting in disaster areas afforded by the Stafford Act. Urgency could be a less compelling reason moving into the recovery phase than it is in response, as expectations for planning and resourcing increase. Second, the decision to award large projects and volumes of projects outside the affected region, for example, seems to ignore the businesses of the region and their need to rebuild their economies; when examining the issue on a state-to-state basis, it becomes clearer that disaster procurement is taken as event-wide, rather than locally focused. Because disaster response begins at the local level, one might reasonably believe that rights and responsibilities should stay at the lowest level of government possible. Federal decision-making may make perfect sense from the perspective of the national government, but for local and state governments, decisions made—even those that are relatively simple, such as contract awards—may fail to ring true. The Stafford Act seems to align with this consideration, with a focus on sending opportunities back to local businesses, but general procurement practice allows for broad discretion that may favor established vendors with national experience, which understand federal procurement processes and know how to win in that environment.

Under the form of urgency and public interest, decisions are made to award very lucrative contracts to firms in patterns that at least suggest a priority for familiarity rather than competitive interest. Because federal rules allow past performance to weigh in decisions, competing interests of social import, namely increasing capacity and reduction of resilience through targeted contracting, perhaps weigh less heavily. Urgency can mean a great many things in a disaster, but it is without doubt that the term is subject to interpretation, and the interpretation may not always yield benefits afforded through legislation. As noted above, the outcomes themselves may show attention to public interest, depending on how one defines the concept. There are assurances in working with long-term partners that may not exist in contracts with newer or less experienced vendors. If that is the public interest, and it may well be, then language of the Act is read as being disingenuous or short-sighted. In this case, from a practical perspective the Act is either ignored or not fully implemented.

Federal awarding authorities appear efficient and effective in targeting very specific firms for large and frequent project awards. The officials responsible for these contracting activities were doing a job and doing it well; it may not have been the job of representing the public interest as envisioned by the Stafford Act. The indifference to the Act's provisions targeting contracting for the impacted area, and frequent use of "other than full and open competition," come together to create a troubling scenario. Alternatively, this may evidence an urging from street-level program administration to revisit the legal authority's language and reflect the realities and challenges of disaster procurement. There could be a need for procurement functions generally to come to terms with the demand for objectivity and transparency while still assuring the public a sound product or service for the money spent (Tadelis 2012).

While there may be some substance to the countering view, the present law is clear about its desire for objectivity and the need for competition, while providing attention to businesses in the affected area. The federal government has a responsibility to provide needed assistance to victims of disaster events, but the proof of its commitment may lie in its approach to contracting. To the extent that agency contracting practices fail to favor businesses in impacted areas, in favor of expediency or even lower cost as a principal determining factor, the aid provided to disaster victims is had at an *unreasonable cost*, in a more enlightened sense of that term, because the Stafford Act as a public law requires an alternative (local-business involvement). A cost that does not afford the victims their fullest chance at resiliency in this region may be too much to bear, if the intent to restore businesses through contract opportunities is indeed sincere. Because the Act does not state when local-business involvement would be less important than, for example, unit cost, we must gather that local-business involvement is important without proviso.

Conclusion

Some important conceptual points with regard to this review deserve our attention. First, the Stafford Act requires attention to the area affected by disaster for contracting, but at least in the case of New Jersey, federal awards to firms in the state as a percent of all awards are not commensurate with the number of contracts occurring in the state or proportional in terms of total value. Success might be defined differently, and it is true that many opportunities related to Hurricane Sandy were awarded to small businesses in the affected area.

The term "small business" as used by the federal government can raise some level of suspicion. Some may see the size standards of small businesses as employed at the federal level to be indicative of businesses that are anything but small. The calculus of federal rules has nevertheless defined small business in a given way,[4] and so outcomes following such guidelines fall along the same lines. Because small businesses may encompass some 90 percent of all businesses in the U.S., as the federal government sees it, then 55 percent of contracts to small business might initially sound successful but be less impressive on closer examination. Even the term "full and open competition" itself is prone to misunderstanding and misapplication. Efforts to encourage a wide range of competitors in such process might be limited by resources available (staff, time and money), or, more objectionably, through nonfeasance in getting the word out about solicitations, or selective notification that tends to favor a more limited contractor pool.

Data reviewed are limited to contracts by federal agencies; we know in disaster contracting that this is only a small part of the total story. There are state and local contracting practices and outcomes with which to contend, but outcomes at these levels are not included on the federal site—

a point the site's introduction admits. However, this poses some concern: if the federal government itself is not particularly good at achieving the broad mandate set out by the Stafford Act then it may be far-fetched to believe that state and local governments have a better understanding for the ideals posed by federal law, and have instituted practices, with more consistency and attention to outcomes, where we may expect a rosy picture at sub-federal levels. The level of understanding across all jurisdictions nationally might be quite uneven, making such a proposition unlikely.

The complexities of federal grants, coupled with the workload of state and local officials in instituting both grant-funded mandates as well as those from their own governmental levels, conspire to thwart strong attempts at compliance with federal provisions. Good faith in implementation may yet yield a failure to comply. Given that there are also those who would seek to avoid following at times competing imperatives, a government of inscrutable rules provides itself no help in achieving favored outcomes.

A second hypothesis posited that transparency in contract reporting is weak. The available data do not tell us the whole story of disaster-related contracting decisions; analyzing data and finding patterns could be difficult prospects for the public, even if interested. From a business perspective, there are considerable opportunities made available in federal contracting for disasters, but there is little doubt of the need for understanding federal contracting in the first place, if a would-be contractor hopes to become successful. As a result, contracting outcomes may be predictable, even where competition is ostensibly welcomed. Contract winners continue to win.

While data assessed here offer a limited foray into reviewing federal procurement data for the purpose of assessing outcomes under the Stafford Act, the effort does highlight some areas in which legal intent and the reality of contracting in a chaotic environment fail to align. As points for additional research, it would be useful to make a more thorough examination of procurement behaviors for low-value contracts (such as those with purchasing cards), to see what sorts of steps are being taken, if any, to bring into being the open opportunity that response and recovery contracting may present for businesses in affected areas. It would also be worthwhile to understand more about how behaviors may change at increasing levels of contract value—not simply in whether or not a solicitation is "full and open," but to determine the extent of efforts made to encourage competition. If other considerations are entering into procurement decisions, even when legal authority allows for preferences intended to increase resilience and decrease vulnerability in the business community, then these considerations may be extraneous and perhaps even costly in ways beyond simple "best product for the price," when rebuilding communities is the goal.

Notes

1 U.S. Census Bureau, American FactFinder, "Government Employment & Payroll." Accessed February 22, 2016. www.census.gov//govs/apes/.
2 FPDS-NG Welcome page. www.fpds.gov/common/html/public_welcome_text.html. It should be noted that the report for Sandy does include a column for purchase card usage; 288 contracts (5.6 percent) indicated "yes" to utilization of a purchase card as the payment method.
3 The FPDS-NG data dictionary defines base and all options value as "the mutually agreed upon total contract or order value including all options (if any). For Indefinite Delivery Vehicles, the estimated value for all orders expected to be placed against the vehicle. For modifications, the change (positive or negative, if any) in the mutually agreed upon total contract value" (Global Computer Enterprises 2015, 32). Hence, values given here include total amounts for base and all options, including negative contract action values as modifications.
4 The U.S. Small Business Administration determines the size of a small business for federal procurement purposes. The SBA utilizes a methodology for determining these standards (www.sba.gov/content/size-standards-methodology). A summary of the standards is located at www.sba.gov/content/summary-size-standards-industry-sector. The full listing of size standards is codified in Title 13 CFR 121. The size standards vary widely—a heavy construction business can make up to $36.5 million in average annual gross receipts, as an example, and still be considered "small."

Bibliography

Atlantic Oceanographic and Meteorological Laboratory. 2014. "The thirty costliest mainland United States tropical cyclones 1900-2013." Accessed February 22, 2016. www.aoml.noaa.gov/hrd/tcfaq/costliesttable.html.

Blake, Eric S., Todd B. Kimberlain, Robert J. Berg, John P. Cangialosi and John L. Beven II. 2013. "Tropical cyclone report: Hurricane Sandy (AL182012)." National Hurricane Center. Last modified February 12. Accessed February 26, 2016. www.nhc.noaa.gov/data/tcr/AL182012_Sandy.pdf.

Daily Mail. 2012. "Revealed: How superstorm Sandy cost New Jersey $30billion in damage to homes, businesses and tourism." *Daily Mail Online*, November 24. Accessed February 22, 2016. www.dailymail.co.uk/news/article-2237731/Superstorm-Sandy-cost-New-Jersey-29-4billion.html.

Dickard, Paul. 2012. "AECOM announced today that it has been awarded an emergency recovery and technical support services contract from the U.S. Department of Homeland Security, Federal Emergency Management Agency (FEMA)." AECOM Press Release. Accessed February 22, 2016. http://tinyurl.com/zxjmf29.

Dreazen, Yochi J. and Jeff D. Opdyke. 2005. "Minorities say Katrina work flows to others." *Wall Street Journal*, September 23, B1.

Esnard, Ann-Margaret and Alka K. Sapat. 2014. *Displaced by Disaster: Recovery and Resilience in a Globalizing World*. New York: Routledge.

Federal Procurement Data System—Next Generation (FPDS-NG). 2016. "Hurricane Sandy report." Accessed January 14, 2016. www.fpds.gov/downloads/top_requests/Hurricane_Sandy_Report.xls.

Garner, Bryan A. 2006. *Black's Law Dictionary, 3rd Pocket Edition*. St. Paul, MN: Thomson West.

Global Computer Enterprises. 2015, orig. 2004. "GSA federal procurement data system—next generation (FPDS-NG) data element dictionary, v. 1.2." Accessed February 22, 2016. www.fpds.gov/downloads/FPDSNG_DataDictionary_1.2.pdf.

Gosier, Chris. 2006. "New reports of Katrina contracting abuse anger lawmakers." *Federal Times*, May 8, 7.

Hayes, Katie. 2014. "Great Lakes reports fourth quarter and year-end financial results." *Business Wire*, February 25. Accessed February 22, 2016. http://investor.gldd.com/releasedetail.cfm?releaseid=828034.

Keyes, W. Noel. 2000. *Government Contracts in a Nutshell, 4th Edition*. St. Paul, MN: Thomson West.

Kildow, Betty A. 2011. *A Supply Chain Management Guide to Business Continuity*. New York: AMACOM.

Lawton, Alan, Julie Rayner, and Karin Lasthuizen. 2013. *Ethics and Management in the Public Sector*. Abingdon: Routledge.

Lehne, Richard. 2006. *Government and Business: American Political Economy in Comparative Perspective, 2nd Edition*. Washington, DC: CQ Press.

National Association of State Procurement Officials (NASPO). 2015. "Non-competitive/sole source procurement: Seven questions." Last modified January 13. Accessed February 26, 2016. www.naspo.org/SoleSourceProcurement/7-Question_Sole_Source_Procurement_briefing_paper-1-13-15.pdf.

Orszag, Peter R. 2009. "Improving Government Acquisition." Last modified July 24. Accessed February 26, 2016. www.washingtonpost.com/wp-srv/politics/documents/Improving_Government_Acquisition.pdf.

Rice, Doyle and Alia E. Dastagir. 2013. "One year after Sandy, 9 devastating facts." *USA Today*, October 29. Accessed February 22, 2016. www.usatoday.com/story/news/nation/2013/10/29/sandy-anniversary-facts-devastation/3305985/.

Smith, Christopher H. 2013. "Floor statement on Sandy supplemental." Last modified January 2. Accessed February 22, 2016. http://chrissmith.house.gov/uploadedfiles/floor_remarks_on_sandy_jan_2_2013.pdf.

Tadelis, Steven. 2012. "Public procurement design: Lessons from the private sector." *International Journal of Industrial Organization*, Vol. 30 No. 3, 297–302.

U.S. Census Bureau. 2016. "County business patterns." Accessed February 22, 2016. http://censtats.census.gov/cgi-bin/cbpnaic/cbpsect.pl.

U.S. Department of Housing and Urban Development, Hurricane Sandy Rebuilding Task Force (HUD). 2013. "Hurricane Sandy Rebuilding Strategy: Stronger Communities, A Resilient Region." Accessed February 22, 2016. http://portal.hud.gov/hudportal/documents/huddoc?id=hsrebuildingstrategy.pdf.

U.S. Government Accountability Office (GAO). 2015. "Disaster contracting: FEMA needs to cohesively manage its workforce and fully address post-Katrina reforms." Accessed February 22, 2016. http://gao.gov/products/GAO-15-783.

3 Revising Federal Disaster Management Policy

Establishing an Officer in Charge

Marc Landy and Jessica Goley

Introduction

During the past decade the U.S. witnessed two natural disasters, Hurricanes Katrina and Sandy, which were so severe as to warrant the title "mega-disaster." Indeed, Katrina inflicted greater economic damage than any other such event in American history. Federal government efforts to assist the response to and the recovery from these dreadful storms met with dramatic success and dismal failure. This chapter examines both the successes and the failures of mega-disaster policy that Katrina and Sandy demonstrated. Based on that assessment, it proposes a revision of the statute that governs federal disaster-management policy in order to redress the glaring weaknesses that Sandy and Katrina exposed.

A crucial aspect of preserving the best of current mega-disaster policy is to avoid excessive centralization of authority, enabling local and state governments to continue to play a critical role. Disaster response poses an especially difficult challenge to American federalism because such response requires speed, efficiency and effective coordination. These qualities are not normally the strong suits of a tripartite federal system encumbered by a welter of duplicative and competing agencies and governing structures. A defense of federalism must acknowledge that a failure to centralize disaster policy-making in the hands of the federal government may indeed result in some human suffering that might otherwise be avoided. However, in addition to providing effective and efficient redress for problems, policies must also be judged on the basis of how well they preserve and protect democratic–republican norms and values. The case for retaining state and local disaster response and recovery initiative fundamentally rests on its essential role in preserving liberty and civic engagement. A central government fully capable of problem-free response to disasters, perhaps by delegating the job solely to the military, would be all too capable of suppressing liberty, ignoring the wants and needs of the affected population as it singlemindedly pursued the orders issued from above. Local initiative and authority is also a vital means for developing and sustaining what Robert Putnam (2000) calls "social capital." If citizens do not believe that they are primarily responsible for working with one another to protect their homes and neighborhoods, then they will not develop the talents and moral fortitude required to preserve and protect a free way of life (Putnam 2000).

Nor does greater centralization always imply increased efficiency and effectiveness. There are important aspects of mega-disaster response and recovery in which states and localities can outperform the national government. Local politicians and officials know their city far better than Washington officials do, and should therefore be able to enable disaster response and recovery to occur in a manner that is far more sensitive to local circumstances and responsive to local concerns. They should also be able to make use of this local knowledge to ensure that disaster-response plans reflect the actual circumstances of life on the ground. Of course, local officials will not always adequately perform this role. Indeed, the City of New Orleans provides a spectacular example of such failure as regards its response to Katrina. Though government is susceptible to inadequacies at all levels, the instance of New Orleans' response to Katrina does not typify local response. Certainly, the dramatic miscalculations and severe deficiencies that comprised New Orleans' response should not detract from the granular knowledge generally held by a government unit most proximal to the citizenry. It is this knowledge that best positions a local government to respond, in contrast to state or federal government.

As the next closest level of government to the localities, states can provide more responsive and sensitive leadership than the national government regarding the inevitable cross-boundary problems that major storms inflict. Also, since the Constitution limits the law-enforcement powers of the national government, states have primary responsibility for backing up the law-enforcement efforts of beleaguered city and county police departments.

In light of these virtues, the authors of this chapter reject those proposals for improving mega-disaster recovery and response that seek to centralize authority in the national government and minimize the contributions and decision-making authority exercised by states and localities.[1] However, the authors do recognize that the very definition of a mega-disaster implies that states and localities will lack sufficient resources to respond and recover solely on their own. They will inevitably have to rely on federal help. As the following analysis makes clear, in the aftermath of both Katrina and Sandy very generous federal help was proffered, and proved invaluable in saving lives, protecting property and enabling significant rebuilding to take place. The critique of the federal role is not based on any lack of generosity but rather on a fundamental flaw in how the role was performed. The revision in the governing statute, the Stafford Act, that we propose is aimed at correcting that fundamental flaw. It recognizes and accepts that because federal aid flows from so many different sources and is managed by so many different agencies, in the absence of aggressive and decisive leadership on the ground, aid will always arrive piecemeal. It will come too slowly and with too many strings attached. Bottlenecks will develop. Spending guidelines will contradict one another. Disputes will fail to be promptly resolved. No amount of statutory tinkering can solve these problems. Plugging one statutory leak will simply result in leaks springing up elsewhere.

The aid that pours into a mega-disaster-affected area is akin to the torrent of soldiers, weapons, tanks and artillery that arrive at a battle scene. The battle cannot be won without them but on their own they cannot prevail. It takes a general to mold these disparate elements into an organized fighting force, lead them into battle and continually adjust and readjust their formations and their tactics as circumstances change and unanticipated problems and opportunities emerge. We propose establishing the mega-disaster equivalent of a general, an Officer in Charge (OIC), who has the authority to command sufficient resources to fight the battle; to invigorate and coordinate the actions of the different fighting forces; to break logjams and to resolve the disputes that will inevitably break out. Since this is a civil and not a military effort, the OIC's discretion must remain far more limited, and the OIC must be subject to far closer congressional scrutiny than a general in the midst of battle. The statutory provisions we propose go to great pains to ensure that the flexibility and initiative to be gained by the creation of an OIC does not entail any sacrifice of congressional authority and oversight.

In one crucial respect the OIC is not like a general, because we also accord this office a critical "peacetime" responsibility. By peacetime we mean the period after the immediate emergency is over, when lives are no longer in danger and essential services have been restored. It is then that attention must turn to recovery—a far more complex political and conceptual task than emergency response. We require the OIC to oversee the development of a strategic recovery plan to be submitted to the president and, if the president approves, to Congress. There is no guarantee that such a plan will adequately preserve the social capital upon which a vibrant recovery depends. But in the absence of such a plan, directed by a person who has been in the thick of the fight, it is far less likely that the proper balance between local initiative and national oversight will be struck.

Section two of the chapter describes the Stafford Act, the statute governing the federal response to disasters. Sections three and four focus on Katrina and Sandy, respectively. This study will show what went right and what went wrong in terms of the overall response to and recovery from these terrible storms. The final section explains why the establishment of an OIC is the best means for addressing the failures revealed in the discussions of Katrina and Sandy.

Review of the Stafford Act

The Stafford Act was passed in 1988 as an amended version of the Disaster Relief Act of 1974. The Stafford Act authorizes the federal government to provide physical and financial assistance during declared emergencies and disasters, and gives FEMA the responsibility for coordinating relief efforts. The Stafford Act was passed to create a systematic and institutional plan for coordinated federal disaster response with state and local authorities (Robert T. Stafford Disaster Relief and Emergency Assistance Act, as amended, and Related Authorities 2007).

The Stafford Act is composed of seven titles. Title I lays out the purpose of the Act and establishes definitions for emergencies and major disasters. Title II authorizes the president's creation of programs for disaster-preparedness, which provide funding and assistance to states for disaster plans and for disaster warning systems. Title III provides for the presidential appointment of coordinating officers in the areas affected by disaster, outlines the responsibilities of the coordinating officer and provides for the formation of emergency support teams composed of federal officials to assist the coordinating officer. It also establishes penalties for misuse of funds, and outlines the requirements for mitigation plans in disaster-preparedness. Title IV lays out the requirements for declarations of major disasters (as opposed to emergencies). It outlines the presidential powers for federal assistance, the essential services that can be provided through federal aid and the proportion of cost that can be assumed by the federal government. It also elaborates the conditions under which Department of Defense resources may be utilized. Title V lays out similar provisions for emergencies (as opposed to major disasters). Title VI lays out requirements for disaster-preparedness, including the provision of resources and training of personnel in anticipation of a disaster. Title VI also explains the authority and responsibilities of FEMA during disasters. These duties include oversight of the preparation of emergency response plans, training and dispersal of funds. Title VII outlines miscellaneous concerns relevant to the Act and its execution. The Act was amended in the Disaster Mitigation Act of 2000. This amendment allows FEMA to impose requirements for mitigation planning as a stipulation of disaster-preparedness grants, and generally expanded the focus on mitigation in disaster planning.

The inadequacies of the federal response to Hurricane Katrina prompted significant changes aimed at improving federal disaster response, including amendments to the Stafford Act. The most significant changes were included in the Post-Katrina Emergency Management Reform Act of 2006. The Post-Katrina Act restructured the Federal Emergency Management Agency (FEMA), providing for new leadership positions with more rigorous qualifications, reexamining the agency's mission and increasing agency authority. These changes were made in response to the prevailing opinion that FEMA's failures during Katrina resulted primarily from the increased emphasis placed on counterterrorism preparedness in the wake of the September 11 attacks, as well as the distribution of disaster-relief responsibilities across the Department of Homeland Security.

In the wake of Hurricane Sandy, additional legislation was passed to further reform and streamline the Stafford Act, reflecting the most significant reforms made to the Act since its initial passage. In January 2013, approximately four months after Sandy made landfall, President Obama signed the combined Disaster Relief Appropriations Act of 2013 and the Sandy Recovery Improvement Act of 2013, together known as the Sandy Supplemental. In addition to providing for these reforms, the Sandy Supplemental created the Federal Transit Administration Emergency Relief Program and provided $50.5 billion for disaster relief to be distributed among the federal agencies involved in disaster recovery. However, no post-Katrina reforms of the Stafford Act have provided for pre-authorized disaster-relief funds to be allocated to federal agencies (U.S. Congress 2013). Had immediate emergency funding been pre-authorized for these agencies, or for distribution by an Officer in Charge, fewer delays would have been suffered as a result of the four-month interval it took for Congress to approve a Sandy relief package.

Hurricane Katrina

Although Katrina was widely considered to reflect the abject failure of government, this is not a fair assessment. In many regards the federal government performed admirably. In Mississippi, within weeks of the onset of the storm, Navy Seabees had cleared more than 200 miles of roads, removed 3,500 tons of debris, delivered 170,000 gallons of water and fuel, repaired more than ninety schools and brought food to more than 600 families a day. The U.S. Coast Guard rescued or evacuated more than 33,500 people and received rave reviews from Gulf coast residents (Ripley 2005). The much-maligned U.S. Army Corps of Engineers (ACE) pumped out 224 billion gallons of water from New Orleans, installed 900 large generators, repaired 170,000 roofs and removed a million cubic yards of debris. The ACE had pre-awarded competitively bid contracts for all of the various response functions to allow quick deployment of resources prior to and immediately after the event (U.S. House of Representatives 2006, p. 217; White House 2006, p. 131). By mid-September the U.S. military had contributed 22,439 active military personnel to build field hospitals, fly helicopter sorties to aid in both evacuation and security, serve meals and transport supplies (American Forces Press Service 2005). As of August 2006 FEMA had provided 101,174 travel trailers and mobile homes as temporary housing for

Hurricane Katrina victims (Federal Emergency Management Agency 2006). If one assumes an average of only three persons per household, FEMA was housing well over a half a million people.

The National Guard performed with distinction. In total, 7,500 troops from four states were on the ground within 24 hours of Katrina (Robbins 2005). As of August 27, the Louisiana National Guard had called almost 3,500 of its members to active duty. By September 9, more than 50,000 guardsmen had been deployed. They performed a variety of missions, including assisting law-enforcement agencies with traffic control and security; transporting and distributing food, water and ice, conducting searches and rescues; providing generator support; and carrying out other missions to protect life and property. On August 28, Louisiana Guardsmen conducted security and screening at the emergency shelter set up at the New Orleans Superdome, where a reported 9,000 to 10,000 local residents reported after heeding the city's mandatory evacuation order issued earlier in the day. As Katrina threatened to flood the low-lying city with water from the Mississippi River and Lake Pontchartrain, further Louisiana Guardsmen set up other shelters and helped state police with evacuations and preparing to support relief operations in the hurricane's aftermath (Miles 2005, p. 202).

Along the Mississippi Gulf, but not in New Orleans, local governments likewise distinguished themselves. For example, in Waveland, Mississippi, policemen were trapped in the station as floodwaters rose. For five hours they clung to bushes in the front yard of the station. When the surge subsided, the officers returned to their duties, not having gone back to their homes or, in some cases, not knowing if they even had a home to go back to.

In Pascagoula, Mississippi, City Manager Kay Johnson Kell said, "If the ox is in the ditch you got to get him out." Because Pascagoula had a rainy-day fund, she could immediately begin debris removal and other tasks without awaiting permissions from federal agencies that were often slow in coming. Because she had negotiated pre-disaster contract, contractors began work immediately at prices stipulated before the storm (Kell 2007).

The fiscal soundness of most Mississippi cities and counties gave them a great advantage in initiating cleanup, but these funds were soon exhausted, with no reimbursement from FEMA in sight. Prompt action by the state of Mississippi enabled the localities to continue their cleanup efforts unabated. By late September the state had created a mechanism to enable localities to borrow money based on the state's moral obligation to repay. Two-year notes were issued with ballooning interest repayment schedules. Deferring interest payments into the future was a recognition that the localities would only be able to make such payments after they began to receive federal reimbursements. Thus the state established a type of bridge financing to give the localities breathing space until federal money became available (Reeves 2008).

Because the Barbour Administration quickly established planning priorities and worked closely both with its own congressional delegation and with HUD, Mississippi was able to begin recovery efforts far more rapidly than Louisiana. It launched its Phase I Homeowners Assistance Program by April of 2006. By November of 2006, over $357 million dollars had been paid to 5,701 homeowners (U.S. Fed News Service 2006). By the end of January 2007, 10,000 homeowners had received checks. As of July 31, 2007, more than $1 billion had been distributed to 13,944 homeowners (Barbour 2007).

Much of what went wrong in Katrina was due to flaws in federal government policy. Three specific programs comprised the heart of the national government's involvement with Katrina. A review of each of them reveals that the key ingredients missing from federal involvement were policy integration and policy authoritativeness. "Integration" means weaving together disparate functionally defined policy threads into a coherent pattern of programs and activities capable of accomplishing the broad purposes that stimulated the call for policy action in the first place. "Authoritativeness" means steadfast adherence to central strategic goals. This often requires an insistence that resources be targeted and concentrated despite powerful political pressures to disperse them more widely. It may also require holding fast to prohibitions and constraints critical to the success of the overall policy design, despite equally powerful political pressures to allow for exceptions, grandfather clauses, waivers or other methods of lessening the impact and hence the effectiveness of those prohibitions and constraints.

In order to establish a designated disaster zone in which investments would benefit from a wide array of tax deductions and credits, GO Zones were created. This concept was modeled after the Liberty Zone established in lower Manhattan in the wake of 9/11. It established a designated disaster zone in which investments would benefit from a wide array of tax deductions and credits. States would also be able to issue new forms of tax-exempt bonds to finance investment in hotels and housing developments, with added incentives for low-income housing, retail outlets, manufacturing facilities and other forms of commercial activity (United States Government Accountability Office 2008, p. 4). The purpose was to stimulate investment in the region.

But, in the absence of a clear priority-setting plan emanating from the executive, Congress did what it is predisposed to do—turn a targeted program intended for a specific category of people and places into a diffuse program benefiting a wider variety of people and places. Senator Lott, a senior member of the Senate Finance Committee, wrote most of the bill's provisions. Instead of focusing on the economic rebirth of storm-ravaged localities and the needs of small business, as the president promised, Lott's economic stimulus package encompassed areas untouched by the storm and favored large investors (GulfCoastNews.com 2005).

The boundaries of the zone extended far beyond the damaged region. Only three counties in Mississippi were hard-hit by the storm, and yet the zone included forty-nine Mississippi counties. In Louisiana only five parishes were hard-hit by Katrina and/or Rita, but nine parishes were included in the GO Zone. In Alabama the damage was confined to two counties near the coast and yet the GO Zone boundary extended eleven counties ranging far to the North, including the thriving university town of Tuscaloosa 200 miles from the coast (United States Government Accountability Office 2008, p. 8).

A 2007 article pointed out that "Investors have long been attracted to college towns and their steady stream of renters. The GO Zone incentives have created an even more enticing reason to consider college towns in areas that qualify for the GO Zone's accelerated depreciation" (Ames 2007). GO Zone investments were evaluated on a first-come, first-serve basis (United States Government Accountability Office 2008, p. 5). This approach greatly favored proposals for undamaged areas whose infrastructure was sound, where insurance was readily available and where prospects for the area's future were rock-solid. It was far more difficult and time-consuming to devise proposals for damaged areas whose infrastructure was still being rebuilt, whose insurance prices were skyrocketing and whose future economic fate was uncertain.

The first-come, first-serve approach also worked well for major national and international corporations with facilities in the damaged areas. Since access to their considerable legal, accounting and capital-market expertise was not affected by the storm, they could quickly mobilize that talent to devise proposals for GO Zone subsidies to repair those facilities (United States Government Accountability Office 2008, p. 31). Indeed, the two largest beneficiaries of GO Zone bonds in Mississippi were Chevron U.S.A and Northrop Grumman Ship Systems, both of whom had large facilities in Jackson County that were severely damaged in the storm. Chevron received $650 million, or about 13 percent of all of Mississippi's GO Zone bonding allocation state's bond authority, and Northrop Grumman received $200 million, more than 3 per cent of the Mississippi allotment (Radelat 2008).

The very decision to frame this the major federally funded economic development plan around bonds and complex tax credit and deduction schemes ensured that it would speak to the needs of large investors rather than small entrepreneurs. The difference between the former and the latter is that the former have access to funds and are trying to determine how best to spend those funds. Hence the attractiveness of tax breaks. The latter do not have sufficient wherewithal to exploit tax subsidies. They need cash. Even if they were in a position to take advantage of the bond financing and tax subsidies the GO Zone offers, they could not afford the teams of lawyers, bankers and accountants needed to structure such complex deals (United States Government Accountability Office 2008, pp. 28–29).

FEMA Public Assistance (PA) funds are used to compensate states and localities for expenditures made for various forms of emergency services and for replacing damaged equipment and rebuilding public infrastructure. This includes police and fire stations, schools, hospitals and other public buildings (Federal Emergency Management Agency 2007). In the case of Katrina, PA proved to be a very burdensome format for aid. It was a *reimbursement* program, which meant that revenue-starved localities had to first find money to pay contractors and then wait to be reimbursed. In practice, localities often failed to pay contractors until the reimbursement money arrived. Knowing that this would be the case, contractors inflated their bids to cover the added debt service they had to pay while waiting to receive their money.

FEMA awarded PA money on a case-by-case basis. Every project—whether it involved removing debris or rebuilding a school building—required a negotiation between the locality letting the contract and the local FEMA official. Routinely, FEMA underestimated the true cost of a project, placing the burden on the locality to once again negotiate the extra costs from FEMA after the project was underway (GAO's Analysis of the Gulf Coast Recovery: A Dialogue on Removing the Obstacles to the Recovery Effort, Testimony of John Thomas Longo 2007, pp. 33–35). According to the Stafford Act, FEMA PA funds could only pay for replacement of preexisting structures. Any improvements had to be paid for by other sources. Because in many cases localities sought to build better schools, courthouses and other buildings, they became mired in endless debates about what aspect of the construction constituted a "replacement," which FEMA should pay for, and what aspect was truly an "improvement," which FEMA would not pay for.

FEMA was permitted to pay for improvements for mitigation purposes, but sometimes it interpreted this loophole narrowly. As an example, if a hospital had 40 out of 100 windows blown out, FEMA would pay to add shutters to protect the windows from future storms for only the forty that had been broken, not for the sixty that had been spared (FEMA's Project Worksheets: Addressing a Prominent Obstacle to Gulf Coast Rebuilding, Testimony of Mark Merritt 2007).

Because FEMA PA was *the* funding source for basic infrastructure—sewage and water pipes, etc.—delays in obtaining funds brought delays in other projects that were infrastructure-dependent. For example, Federal Highway Administration funds were available to repave many of the major streets in New Orleans, but those streets sat above a crumbling infrastructure of sewage and water pipes that were deteriorating even before the storm, but had then been corroded by weeks of exposure to salt water.

In late December Congress responded to requests from both the Mississippi and Louisiana delegations for additional funds and for more flexibility in how funding could be spent. Congress appropriated part of the 2006 Defense spending bill (HR 2863), including funds originally granted to FEMA but now made available to the states in a more desirable discretionary fashion of Community Development Block Grants (CDBG) (Harris 2006).

CDBG are the federal government's largest and most widely available source of financial assistance to support state and local government-directed neighborhood revitalization, housing rehabilitation and economic development activities—precisely the set of activities that communities devastated by Katrina required to recover. The block-grant nature of the program afforded local and state officials a great deal of discretion in determining which combination of the twenty-five categories of eligible activities to undertake when developing their community development plans. Eligible CDBG activities include historic preservation; real property acquisition, demolition, site preparation and disposition; economic development and job creation, including assistance to for-profit entities and establishment of revolving loan funds; housing assistance, including rehabilitation loans and grants; public service activities, including job counseling and employment training; and assistance to not-for-profit entities, including community development corporations and faith-based institutions.

Under normal circumstances, in order for an activity to be eligible for CDBG funding it had to principally benefit low- and moderate-income persons; aid in eliminating or preventing slums or blight; or meet particularly urgent community development needs because existing conditions posed a serious and immediate threat to the public. Helping storm-ravaged communities to recover clearly met the third objective (U.S. Department of Housing and Urban Development 2014). Unlike FEMA money, CDBG money could be spent very freely, but the application process for obtaining the funds proved highly complex, slow and burdensome. In the words of Andy Koppelin, former director of the Louisiana Recovery Authority, the CDBG application process is "a maze but there is cheese at the end. FEMA just tells you, 'You can't'" (Koppelin 2006). Among the obstacles created by the CDBG application process was the need to show compliance with the Davis–Bacon Act that requires contractors to pay prevailing wages and, in order to prove that there would be no duplication of benefits, the deduction of all forms of other assistance, including private insurance, from the CDBG payout. This requirement was particularly onerous because of the lack of adequate databases for checking on the various forms of federal expenditures and the unwillingness of harried insurance companies to redeploy staff away from dealing with irate customers to gather the requisite data.

The CDBG application's capacity to slow down recovery is illustrated by the onerous requirements it placed on the Road Home, Louisiana's program for aiding homeowners whose houses had been damaged or destroyed by Katrina. The Road Home required its grantees to rehabilitate their properties rather than simply take the grant money elsewhere. It put a lien on their property to be removed only after the rehabilitation had fully taken place. The fact that this requirement had anti-mitigation implications did not trouble HUD. Rather, it insisted that CDBG rules required each lien trigger a separate Environmental Impact Statement (EIS). Requiring every homeowner to go through the onerous and costly soil and water sampling which an EIS involves would have paralyzed the program. To avoid triggering the EIS, the LRA abandoned the lien and substituted a promise from the grantee that he/she would occupy the property for three years. After time-consuming negotiations, HUD agreed that such a promise would not require an EIS. Then it reversed itself and once more demanded grantee-by-grantee EIS's. More time passed and then the LRA proposed another form of EIS evasion. The LRA would not itself demand evidence that rehabilitation was progressing; rather, it would create a separate entity which would issue the grant and which would hold the Road Home grant money in escrow, doling it out to the homeowner as the repair proceeded. HUD finally accepted this proposal, removing a critical obstacle to the commencement of what was supposed to be a crash program enabling homeowners to escape from their cramped FEMA-provided trailers and move back to their homes (Leger 2007, pp. 6, 17).

The extraordinary paperwork burdens and implementation snafus that marked the distribution of federal recovery aid mask a deeper problem that the CDBG program posed, particularly in Louisiana. Unlike Mississippi, which reserved a significant fraction of its CDBG funding for redeveloping the port of Gulfport and building infrastructure north of I-10, Louisiana devoted almost all of its CDBG funds to housing. Louisiana's Road Home entitled homeowners to a maximum of $150,000 to use either to rebuild their homes or to purchase homes elsewhere in the state. The percentage of the grant they received depended on the percentage of damage done to their home. All other forms of compensation—including other federal and state grants, as well as flood and homeowners' insurance—were deducted from the Road Home grant, meaning that those receiving more than $150,000 from insurance were not eligible and that $150,000 was an absolute cap on total compensation for anyone receiving Road Home assistance.

As the name implies, the purpose of the program was to bring people back to Louisiana. If homeowners chose to use the money to purchase a home outside the state, they received 40 percent less than if they chose one of the two other options. And the unintended consequence of the $150,000 cap was that the flexibility of home-siting would be further limited by a lack of sufficient funds to buy or build homes in less flood-prone areas of the city or state. $150,000 was simply not enough money to rebuild anywhere else than at the site of the damage, and the fractional basis for calculating the grant meant that many would receive far less than $150,000. The Road Home thus provided a powerful incentive for homeowners to refurbish their existing properties on their existing footprint.

The existence of an OIC, as described later on, would not have remedied all the weaknesses of these three programs, although it would have been able to cut through much of the red tape they created. Rather, by presenting Congress with a well-designed, targeted recovery plan that it could only vote up or down, it could help Congress to avoid the distributional weakness to which it is prone, and to provide means for overcoming the coordination problems that statutory stove-piping creates.

Hurricane Sandy

Hurricane Sandy was the first major natural disaster that tested the effectiveness of the improvements made to federal emergency management following Hurricane Katrina. Sandy resulted in the damage or destruction of at least 650,000 homes, and caused at least $50 billion in damages (Blake *et al.* 2013). Although the damage inflicted by Hurricane Sandy was less severe than that inflicted by Katrina, it affected a larger swathe of territory. The federal government issued federal disaster declarations in twelve states and the District of Columbia, and the most serious damage occurred across an approximately 140-mile stretch of densely populated coastline in New Jersey, New York and Connecticut. Additionally, major damage sustained across the New York metropolitan area—the largest metropolitan area in the U.S.—created extraordinary recovery challenges (Hurricane Sandy Rebuilding Task Force 2013, pp. 18–22).

Because of the number of states, local governments, major metropolitan areas and vital transit systems affected, Hurricane Sandy can be considered to have posed an even greater challenge for the federal component of emergency management than did Hurricane Katrina, despite being a less severe storm. Peter Rogoff, Administrator of the Federal Transit Administration, called Hurricane Sandy "the worst public transit disaster in U.S. History," with it rendering more than half of all transit trips in the U.S. unavailable on the day of the storm (U.S. Congress 2013, p. 3). However, because of the profound damage to interstate transit systems that fall under the jurisdiction of the Federal Transit Administration, Federal Highway Administration and Federal Railroad Administration, in some areas the federal government's role in the recovery fell clearly under the purview of certain agencies and was therefore subject to fewer institutional challenges.

The actions taken in the wake of Sandy which came closest to providing the leadership necessary for effective recovery and coordination across levels of government were encompassed in Executive Order 13632, issued by President Obama on December 7, 2012. This order established the Hurricane Sandy Rebuilding Task Force, a temporary task force responsible for coordinating interagency participation in the federal rebuilding process. The Task Force was composed of members from twenty-seven federal agencies and the White House, and was primarily responsible for developing the Hurricane Sandy Rebuilding Strategy. The Hurricane Sandy Rebuilding Strategy formulated an approach for long-term rebuilding and recovery, and included sixty-nine recommendations in eight policy areas (Brown 2014). However, the Task Force was designed mainly for the purpose of communication between federal governmental departments, and lacked the authority to grant emergency funding or resolve interagency disputes (Executive Order 13636: Establishing the Hurricane Sandy Rebuilding Task Force 2012).

In many ways, the response to Hurricane Sandy was more successful than was the response to Katrina. The states and municipalities affected by Sandy displayed a far greater degree of effectiveness in responding to the emergency than the City of New Orleans did with Katrina. The Mississippi Gulf Coast towns and counties hit by Katrina and the State of Mississippi did a much better job, but their fine performance did not erase the sense of utter incompetence instilled by the media's fixation on New Orleans.[2]

Hurricane Katrina was responsible for more than 1,800 deaths, mainly due to failures to develop and maintain levees and floodgates necessary to protect the City of New Orleans and inadequate evacuation planning and enforcement in the days before the storm (Graumann *et al.* 2005). Incomplete evacuation in the most severely affected areas dramatically increased the danger and duration of search and rescue operations, delayed the restoration of vital services and contributed to the infamous overcrowding at the Superdome and Convention Center and the breakdown of law and order across the city.

By contrast, evacuation orders were issued with greater notice and were more effectively executed prior to Hurricane Sandy. This made it easier for the focus of state and local governments to turn more quickly to restoration of vital services. Although more than 8.5 million customers lost power as a result of Hurricane Sandy, and recovery was complicated by a "Nor'easter" that struck the same area nine days after the hurricane, service was restored to most customers within three weeks of the storm (U.S. Department of Energy 2012). New York City experienced extensive flooding and power outages that ground the transit system and business to a halt, but conditions never came close to the dire circumstances in New

Orleans after Katrina. The Bloomberg administration implemented a program called Rapid Repairs tasked with restoring essential services to as many damaged buildings and homes as quickly as possible, in order to allow people to "shelter in place" rather than be housed in trailers, hotels and emergency relief shelters provided by FEMA. In total $640 million was spent to restore services to nearly 12,000 buildings in less than 100 days (Buettner and Chen 2014).

The greatest shortcomings of the federal approach to Hurricane Sandy have been in the planning and implementation of long-term recovery, rebuilding and mitigation efforts. In these areas the response has suffered from many of the same problems as the Katrina response. Application processes for individuals were difficult to navigate, filled with redundancies, and demanded verification materials whose provision was often impossible for families whose homes had flooded. Rebuilding efforts failed to adequately consider and implement non-structural alternatives for mitigation. Despite extensive and protracted oversight measures to prevent improper payments like those that beset the Katrina recovery, many applicants who received payments to cover the cost of rebuilding measures have been asked to return funds that were improperly granted. These problems resulted from a recovery plan mired in bureaucratic inefficiencies and redundancies, and a lack of clear and consistent leadership.

The Build it Back program implemented in New York City, one of the areas hardest hit by Sandy, shows in high relief the systemic problems that plagued the recovery more generally. Many of the program's shortcomings stem from features designed to avoid problems that plagued the Katrina recovery. In order to prevent corruption, instead of issuing payments to homeowners, the New York program was designed such that contractors would be directly hired by the program with federal funds for repairs on private homes. The application process was designed to avoid improper payments that would later have to be rescinded. Repairs to qualifying homes were focused on long-term investments rather than temporary solutions, and low-income applicants received priority.

However, these efforts to prevent the problems that impaired the Katrina recovery failed on most fronts because of bureaucratic slowdown and a lack of effective leadership. The administration of the recovery program in New York City illustrates the broader problem. The city received federal funding to create its own recovery program with very little oversight. The Bloomberg administration hired Boston Consulting Group to design the Build it Back program at a cost of more than $8 million for the first six months of work. Boston Consulting Group hired Public Financial Management, a company that had never previously managed disaster recovery, to oversee the program. Public Financial Management hired URS Corporation to handle the intake centers, and URS in turn hired untrained temporary workers to staff the centers and designed software to administer the application that arrived late and was fraught with errors. The program underwent three leadership changes in its first year. It took eight months for the city to open application centers following the storm. The application process had so many steps and required so much verification that it was nearly impossible to reach the approval stage. Families were called in to provide the same materials multiple times because the application software frequently lost necessary documents. Because the program prioritized low-income applicants, who often had greater difficulty providing the materials required by the rigid application system, middle-class homeowners who made up the vast majority of applicants experienced prolonged delays. Nearly two years after the storm, construction had not commenced on the homes of any of the 20,000 homeowners who had applied for recovery aid. As of September 2014, nearly 13,000 people still awaited assistance, and 6,000 homeowners had abandoned their applications (Buettner and Chen 2014). Despite bureaucratic delays being justified on the understanding that they would prevent improper payments, as of September 2014 FEMA was reviewing payments issued to 4,500 homeowners and had requested the return of $5.8 million in federal aid (Caruso and Kunzelman 2014).

Although the New York City program is an isolated example of the broader recovery—and a particularly extreme example, insofar as its response has been the most highly criticized—it is indicative of broader efficiency and organizational problems that face local, state and federal government agencies in emergency response. In New Jersey, although the response has been more successful, two thirds of applicants to the main housing recovery program still awaited approval and payments in October 2014. Although the Christie administration has taken steps to reform the process, like in New York, these delays are attributed to "overly rigid rules, an opaque application process, and too much bureaucratic red tape" (Gurian 2014).

Efficiency is only one measure of the success of the response to a major natural disaster. Important considerations also include the extent to which the response incorporates mitigation efforts and non-structural rebuilding alternatives to prevent recurrence of the same damage in future severe weather events. The Hurricane Katrina response was particularly lacking in this area. Some measures have been taken to prevent future damage and rebuilding to the same high-risk areas since Katrina, but they have been limited and have faced significant pushback. The Biggert–Waters Flood Insurance Reform Act of 2012 began the process of phasing out flood insurance subsidies in high-risk flood zones (Knowles and Kunrether 2014). These subsidies discouraged people from relocating to areas with safer elevation. The Biggert–Waters reforms took important steps to prevent the same damage and expenditures from recurring in future storms, but they have been plagued by setbacks. The Consolidated

Appropriations Act of 2014 and the Homeowner Flood Insurance Affordability Act of 2014 repealed and prohibited the implementation of crucial sections of the Biggert–Waters Act in the wake of resistance from constituencies in high-risk areas. Even when successful, these measures do not go as far as is necessary to mitigate future damage and create efficiency in disaster recovery spending (Loudin 2013). According to Lieutenant General Boswick, Commanding General and Chief of Engineers of the Army Corps of Engineers, no non-structural alternatives for mitigation had been requested or implemented as of a year after the storm (U.S. Congress 2013).

The failures of the Hurricane Sandy recovery did not result from a lack of good intentions. Genuine efforts to make improvements to federal emergency management in light of lessons learned from Katrina failed because of a lack of leadership, bureaucratic coordination and pre-authorized funding. Such efforts would have been rendered more effective by an appointment to a leadership position vested with the power and authority to coordinate, prioritize and streamline recovery and mitigation efforts following a national mega-disaster.

An Officer in Charge?

The chief reason for enabling the president to appoint an Officer in Charge (OIC) is that such a person can provide the leadership necessary to promote policy integration and policy authoritativeness in the face of the inevitable bureaucratic and congressional obstacles to recovery from mega-disasters. Congress should provide the president with standby authority to appoint such a person immediately after a mega-disaster occurs. Each OIC would be appointed under a specific executive order unique to the mega-disaster at hand. This is necessary because the precise limits and extent of the OIC's authority need to be tailored to the specific physical political and economic circumstances involved. The OIC would be appointed for a limited term sufficient for him to perform the functions detailed below, and no longer. Because the political role which the OIC must perform requires that they enjoy enormous prestige both in the affected region and with Congress, it is not a job for a mid-level functionary or even for the head of a federal agency such as FEMA. It might perhaps go to a prominent cabinet official, but only if that official were to take full leave of his or her existing responsibilities. A more likely choice would be a prestigious former high-level public servant. In 2005, Colin Powell would have been an estimable choice.

The Officer in Charge would perform three essential tasks: (1) work with officials from FEMA, the Coast Guard and other relevant federal agencies to coordinate the federal response effort and take the lead in resolving interagency disputes and conflicts with state and local emergency responders; (2) convene deliberations among the relevant federal, state and local actors regarding recovery planning; (3) determine whether or not to propose a national recovery plan to the president and, if a plan is called for, what that plan should contain. In that event, the OIC would be required to submit such a plan to the president within six months of the OIC's appointment.

As we have demonstrated, with the exception of some instances in New Orleans, local first responders did a magnificent job in both storms. Therefore, the presumption of local and state initiative regarding first response should not be and is not undermined by the OIC proposals. Had an OIC been in place in Katrina, the OIC response role in Mississippi would have been limited to expediting logjams resulting from jurisdictional dispute and interagency quarrels. However, as the New Orleans example shows, there is a possibility that a full-fledged breakdown of local authority can occur in the midst of a mega-disaster. Therefore the OIC statute should provide for an emergency declaration of federal receivership in the event of such a case of governmental collapse. This would have enabled the OIC to take command of the first response. Because the instance of New Orleans after Hurricane Katrina is extreme, the OIC statute would be written to set a high bar for initiating such a drastic usurpation of local authority.

As both Sandy and Katrina demonstrated, the bulk of the problems involving the federal role in disaster management occurred at the recovery, not the response, phase, and it is in the recovery phase that the OIC would prove most helpful.[3] The OIC would have the freedom to ensure that all relevant parties are brought to the table to consider policy option and policy design. The OIC's efforts to coordinate response will have sensitized the OIC to the debilitating effects of stove-piping and therefore make them especially sensitive to the need for policy integration. The most important attribute of the OIC Plan would be its authority to integrate different policy tools, implementation practices and agency personnel to maximize the likelihood that recovery efforts will build on rather than detract from one another. Thus if one agency insists on the use of affirmative action as a prerequisite for funding and another agency forbids it, the OIC would have the authority to cut that Gordian knot. Because the OIC Plan is not a negotiated product but rather the fruit of a single strategic vision with full control over policy design architecture, it can embody the resoluteness of purpose and the commitment to priority-setting that a robust recovery requires.

If the OIC submits a plan to the president and the president decides to submit it to Congress, Congress would subject the specific policy proposals contained in the report to a fast-track procedure. This procedure would resemble the one Congress has adopted for international trade agreements and would be in the same spirit as its approach to the proposals of the various Base Closing Commissions. Congress would agree to bring the Report to the floor within forty-five days, have a floor vote within ten days, ban all amendments and, with regard to the Senate, ban the filibuster. The congressional vote would be either to accept or reject the proposal in its totality.

It is a great deal to ask of Congress to voluntarily limit its legislative discretion in this manner. But nine months would have already passed since the mega-disaster and the need to commence actual recovery operations would be urgent. There would no longer be time for the slow process of subcommittee and committee hearings and mark-ups that are the congressional norm. The OIC's proposals might prove insufficiently coherent and integrative, but that problem is unlikely to be fixed via piecemeal alteration and amendment.

The OIC's report will explain and defend its overall strategic rationale, thereby enabling Congress to have a serious debate about the merits of the proposals. The report is intended to enable Congress to evaluate the overall goals of the recovery proposals, to decide whether it supports those goals and to assess whether the specific proposals are sufficiently comprehensive, integrated and practical to accomplish those goals. If Congress disagrees with the goals of the report or doubts the efficacy of its proposals, it can and should vote the proposals down.

The fast-track approach will have the added virtue of disciplining the OIC. It deprives the OIC of the ability to defer the hardest choices to Congress. Therefore, it improves the likelihood that the report's recommendations will prove sufficiently tough-minded and comprehensive to warrant congressional approval.

Although there are obvious problems with stretching recovery deliberations over nine months, there are advantages as well. In the immediate aftermath of a mega-disaster it is difficult to engage in the dispassionate thinking and discussion needed to consider how best to re-invest. There is a need for a period in which the immediate and most desperate needs of victims are attended to before sufficient regard can be given to the facts of the situation and to the hard choices they impose for the future. If Congress approves the OIC's recommendations, then an equally difficult stage, that of implementation, commences. If the president determines that the OIC has excelled during the process of formulating the recovery plan, he should have the discretion to extend the OIC's term of service.

The OIC is not a czar, although it is a comparable role. The term "czar" has been applied to many diverse efforts by presidents to overcome bureaucratic wrangling or otherwise treat problems that the existing bureaucratic chain of command seems incapable of adequately addressing. President Obama has appointed czars for such diverse purposes as establishing and enforcing pay guidelines for executives of companies receiving federal bailout monies; overseeing the bailout and rescue of the automobile industry; coordinating the cleanup of the Great Lakes; and promoting "green" jobs. In their book *The President's Czars*, Mitchel A. Sollenberger and Mark J. Rozell provide a comprehensive consideration of the czar phenomenon. They define a czar who operates in the arena of domestic policy as:

> An executive branch official who is not confirmed by the Senate and is exercising final decision making authority that often entails controlling budgetary programs, administering/coordinating a policy area, or otherwise promulgating rules, regulations and orders that bind either government officials and/or the private sector.
> (2012, p. 7)

Sollenberger and Rozell offer trenchant criticisms regarding the deployment of czars. Most importantly, they demonstrate that czars violate the separation of powers, since they exert powers not properly delegated to them; operate without congressional oversight; mimic the role and responsibilities of cabinet officers; and spend money that Congress has not authorized them to spend.
(2012, pp. 6–8, 17)

We agree with these criticisms and therefore we must point out the ways in which the OIC differs from Sollenberger and Rozell's critique. First of all, the OIC's duties and responsibilities *are* defined by Congress, in the Act which grants the president the standby authority to appoint one. Second, congressional oversight is stringent in that the OIC's plan can only go into effect if it is approved by Congress. Third, as discussed in the next section, the Act establishing the OIC clearly specifies the conditions under which an OIC may be called into existence. Indeed, the OIC fits the Sollenberger and Rozell definition of a czar only in that the OIC is not confirmed by Congress and does coordinate policy. In and of themselves, these attributes are not constitutionally problematic. The power of the OIC to cut through red tape is limited to agency regulations. The OIC has no power to overrule or ignore acts of Congress.

The OIC is hardly unique as a device for policy coordination and planning among diverse federal agencies. The Joint Chiefs of Staff, the National Security Council and the Office of National Intelligence are three prominent examples of such an effort. The OIC differs from them, however, in that the OIC is activated in response to a specific emergency and is obligated to submit one single plan. Upon implementation of this plan, the OIC no longer serves in this role.

Unless there are stringent standards for defining a mega-disaster, the Officer in Charge approach risks becoming a new form of moral hazard. Governors and state congressional delegations will be tempted to pressure the president to declare their serious hurricane, fire or blizzard to be a mega-disaster with the expectation that this designation will produce more federal aid than could be obtained through ordinary channels. This concern is not merely hypothetical. During the Clinton years the criteria for obtaining federal disaster relief were loosened and, despite a relative lack of truly major incidents in this period, the number of disaster declarations and the amount of disaster-relief spending rose significantly (Roberts 2006).[4] Hurricanes are not the only significant forms of serious natural disaster, but in the past many decades they have produced catastrophic levels of death and property damage that far exceed their rivals—volcano eruptions, wildfires and earthquakes. Hurricane Katrina inflicted an estimated $81 billion in damage. This figure is nearly twice that of the damage assessment for Hurricane Andrew, the worst prior hurricane, which inflicted approximately $43 billion in damage (Roberts 2006). Katrina killed at least 1,300 people (Associated Press 2006). By comparison, the deadliest and costliest earthquake, the one that occurred in Northridge, California in 1994, killed sixty-three people. Estimates of property damage range between 13 and 20 billion dollars (U.S. Geological Survey 2012). As its damage exists somewhat centrally to other incidents, Hurricane Andrew should serve as the benchmark for designation of a mega-disaster. If a storm is assessed as inflicting more than 43 billion dollars' damage, in constant dollars, an Officer in Charge would be appointed by the president.[5] To facilitate this, we suggest that an interagency damage-assessment process be established for hurricanes, probably under the auspices of the National Hurricane Center. It would be activated in advance of an oncoming large hurricane which might be expected to reach Category 3 level or higher. The U.S. Geological Survey should also be part of the process so that in the case of a mega-disaster earthquake or volcanic eruption, persons capable of providing expert damage assessments would be prepared to do so.

Conclusion

The aim of this statutory proposal is to address the problem that emergencies create for the rule of law. On the one hand, the crisis posed by a mega-disaster cannot be adequately dealt with by existing federal programs, as those are normally implemented by many agencies on an *ad hoc* basis, and too slowly. They are too reactive and uncoordinated. Efforts at statutory fixes of the specific weaknesses may lead to some improvement but no amount of tinkering can obviate the need for flexible, authoritative leadership on the ground. The OIC proposal is designed to establish leadership of the necessary caliber without undermining congressional oversight and congressional supremacy. In the immediate aftermath of the crisis, the OIC would enjoy considerable decision-making freedom. But the OIC only comes into existence under very extraordinary circumstances: a mega-disaster, as defined above. And the OIC's high degree of discretion ends when the response phase, a short period of time, ends. The OIC's more important duty is to provide planning leadership during the subsequent recovery phase. But the OIC has no power to implement a plan. Congress decides whether or not the OIC's plan is to go into action. The OIC is the instigator and the mobilizer of deliberation, but Congress still retains the legislative authority granted to it by the Constitution.

Notes

1 For a review of such centralizing proposals see Marc Landy, "Review Essay: A Failure of Initiative: Final Report of the Select Bipartisan Committee to Investigate the Preparation for and Response to Hurricane Katrina and The Federal Response to Hurricane Katrina Lessons Learned." *Publius: The Journal of Federalism* 38, No. 1 (2008): 152–165.
2 For a detailed comparison of the response of New Orleans, Louisiana, the towns along the Mississippi Gulf Coast and Mississippi see Marc Landy. "Mega-Disasters and Federalism." *Public Administration Review* 68, Supplement to Volume 68: The Winter Commission Report Revisited: 21st Century Challenges Confronting State and Local Governance and How Performance Can Be Improved (2008): S186–S198.
3 In response to Hurricane Andrew, President George H. W. Bush appointed Andrew Card to serve as a coordinator of the interagency taskforce created to deal manage the storm response. Students of the response effort vary in their evaluation of Card's performance in that role. See Saundra K. Schneider. *Flirting with Disaster: Public Management in Crisis Situations* (Armonk, NY: M.E. Sharpe Inc., 1995), 95 and Gary L. Wamsley and Aaron D. Schroeder, "Escalating in a Quagmire: The Changing Dynamics of the Emergency Management Policy Subsystem." *Public Administration Review* 56, No. 3 (1996): 235–244.

In any event, his role did not involve the planning leadership we ascribe to the OIC. Similarly, George W. Bush appointed a coordinator, Donald Powell, to improve the federal response to Katrina, but with little of the authority that we ascribe to the OIC. Like Card, he was not tasked to submit a plan for recovery. See Martha Carr, "Powell to Step Down," *NOLA.Com/The Times Picayune*, February 29, 2008, www.nola.com/news/index.ssf/2008/02/donald_powell_to_step_down.html
4 In addition read Richard Sylves and Zoltán I. Búzás, "Presidential Disaster Declaration Decisions, 1953–2003: What Influences Odds of Approval?". *State & Local Government Review* 39, No. 1 (2007): 3–15. www.jstor.org/stable/4355437 and also Mary W. Downton and Roger A. Pielke Jr., "Discretion without Accountability: Politics, Flood Damage, and Climate." *Natural Hazards Review* 2, No. 4 (2001): 157–166.
5 Since no one triggering metric is adequate under all circumstances, substitute measures of damage should be permitted. For example, if the dollar damages that a mega-disaster causes in a low-income region do not reach or exceed $40 billion because of pervasive low property values, alternative damage measures, such as number of homes destroyed or number of persons displaced, could be incorporated into the metric.

Bibliography

American Forces Press Service. 2005. "Military Continues Hurricane Katrina Support." http://archive.defense.gov/news/newsarticle.aspx?id=17316. Accessed December 4, 2014.

Ames, Eric. 2007. "Top 5 GO Zone College Towns." *NuWire Investor.* June 1. www.nuwireinvestor.com/articles/top-5-go-zone-college-towns-51085.aspx. Accessed December 5, 2014.

Associated Press. 2006. "Death Toll From Katrina Likely Higher Than 1,300." New Orleans, February 10. www.nbcnews.com/id/11281267#.VvfhFL4rKM8. Accessed October 10, 2014.

Barbour, Haley. 2007. "Homeowners Grant Program Passes $1 Billion Mark." Jackson, MS, August 1. http://votesmart.org/public-statement/282625/governor-barbour-homeowners-grant-program-passes-1-billion-mark#.VvaeCb4rKM8. Accessed December 6, 2014.

Blake, Eric S., Robert J. Kimberlain, John P. Cangialosi, and John L. Beven. 2013. "Tropical Cyclone Report." National Hurricane Center. www.nhc.noaa.gov/data/tcr/AL182012_Sandy.pdf. Accessed November 15, 2014.

Brown, Jared T. 2014. "The Hurricane Sandy Rebuilding Strategy: In Brief." Congressional Resesarch Service. www.gpo.gov/fdsys/pkg/FR-2012-12-14/pdf/2012-30310.pdf. Accessed October 4, 2014.

Broussard, Ernest Jr. 2007. "GAO's Analysis of the Gulf Coast Recovery: A Dialogue on Removing the Obstacles to the Recovery Effort, Testimony." *Hearing before the Ad Hoc Subcommitteee on Disaster Recovery of the US Senate Committee on Homeland Security and Governmental Affairs.* 110th Congress, First Session, April 12. www.gpo.gov/fdsys/pkg/CHRG-110shrg35524/html/CHRG-110shrg35524.htm. Accessed September 9, 2014.

Buettner, Russ, and David W. Chen. 2014. "Hurricane Sandy Recovery Program in New York was Mired by Its Design." *New York Times*, September 4. http://nyti.ms/1nyANWW. Accessed October 1, 2014.

Carr, Martha. "Powell to Step Down." *NOLA.Com/The Times Picayune*, February 29, 2008. www.nola.com/news/index.ssf/2008/02/donald_powell_to_step_down.html. Accessed November 30, 2014.

Caruso, David B., and Michael Kunzelman. 2014. "FEMA Tells Residents it Needs Sandy Money Back after Wrongly Dispursing Funds." *Washington Post*, November 9. www.washingtonpost.com/politics/fema-tells-residents-it-needs-sandy-money-back-after-wrongly-disbursing-funds/2014/11/09/bed1233a-685a-11e4-a31c-77759fc1eacc_story.html. Accessed September 14, 2014.

"Executive Order 13636: Establishing the Hurricane Sandy Rebuilding Task Force." 2012. Vol. 77. December 14. www.gpo.gov/fdsys/pkg/FR-2012-12-14/pdf/2012-30310.pdf. Accessed September 11, 2014.

Federal Emergency Management Agency. 2006. "Draft White Paper Wind Mitigation Recommendations for FEMA Travel Trailers." www.fema.gov/media-library-data/20130726-1724-25045-6789/wind_mitigation_recommendations_for_fema_travel_trailers.pdf. Accessed August 11, 2016.

Federal Emergency Management Agency. 2007. "Public Assistance Guide." June. www.fema.gov/public-assistance-policy-and-guidance/public-assistance-guide. Accessed November 15, 2014.

Graumann, Axel, Tamara Houston, Jay Lawrimore, David Levinson, Neal Lott, Sam McCown, Scott Stephens, and David Wuertz. 2005. *Hurricane Katrina: A Climatological Perspective—Preliminary Report*. National Climactic Data Center. www.ncdc.noaa.gov/oa/reports/tech-report-200501z.pdf. Accessed November 16, 2014.

GulfCoastNews.com. 2005. "Lott Joins President Bush at White House for Signing of Senator Lott's Katrina Tax Incentive Bill." December 21. www.gulfcoastnews.com/GCNarchive/2007/GNCnewsKatrinaLottGoZone.htm. Accessed December 10, 2014.

GulfCoastNews.com. 2006. "HUD Approves Homeowner Assistance Program, Application Process Begins April 17." April 6, 2006. www.gulfcoastnews.com/GCNarchive/2006-2005/GCNnewsKatrinaHomeownerPrgApproved.htm. Accessed August 11, 2016.

Gurian, Scott. 2014. "Two Years After Hurricane Sandy, New Jersey's Recovery Trudges Along." *New Jersey Spotlight*, October 29. www.njspotlight.com/stories/14/10/29/two-years-after-hurricane-sandy-new-jersey-s-recovery-trudges-along/. Accessed December 10, 2014.

Harris, Kathleen. 2006. "Harris Commends Secretary Jackson for Providing Hurricane Assistance." *Vote Smart*. January 27. https://votesmart.org/public-statement/150649/harris-commends-secretary-jackson-for-providing-hurricane-assistance#.VugdXfkrKM8. Accessed December 10, 2014.

Higa, Liriel. 2005. "Legislative Summary: Hurricane Supplementals and Related Aid." *CQ Weekly*, January 2: 22.

Hurricane Sandy Rebuilding Task Force. 2013. "Hurricane Sandy Rebuilding Strategy." 18–22. http://portal.hud.gov/hudportal/documents/huddoc?id=HSRebuildingStrategy.pdf. Accessed August 11, 2016.

Kell, Kay Johnson, interview by Marc Landy. 2007. *City Manager of Pascagoula* (March 6).

Knowles, Scott Gabriel, and Howard Kunrether. 2014. "Troubled Waters: The National Flood Insurance Program in Historical Perspective." *Journal of Policy History*, July: 327–353.

Koppelin, Andy, interview by Marc Landy. 2006. *Executive Director of the Louisiana Recovery Authority*. March 8.

Leger, Walter. 2007. "Written Testimony of Walter Leger, Member of the Board of the Louisiana Recovery Authority, before the U.S. Senate Subcommittee on Response and Recovery." January 29.

Longo, John Thomas. 2007. "GAO's Analysis of the Gulf Coast Recovery: A Dialogue on Removing the Obstacles to the Recovery Effort, Testimony." *Hearing before the Ad Hoc Subcommitteee on Disaster Recovery of the US Senate Committee on Homeland Security and Governmental Affairs*. New Orleans, LA: 110th Congress, First Session, April 12. 33–35. www.gpo.gov/fdsys/pkg/CHRG-110shrg35524/html/CHRG-110shrg35524.htm. Accessed October 13, 2013.

Loudin, L. Keith. 2013. "The Biggert–Waters Flood Insurance Reform Act of 2012: Elevation Ratings." Federal Emergency Management Agency. www.fema.gov/media-library/assets/videos/84705. Accessed October 16, 2014.

Merritt, Mark. 2007. "FEMA's Project Worksheets: Addressing a Prominent Obstacle to Gulf Coast Rebuilding, Testimony." *Ad Hoc Subcommittee on*

Disaster Recovery of the Committee on Homeland Security and Governmental Affairs. 110th Congress, July 10. http://catalog.hathitrust.org/Record/005844461. Accessed December 4, 2014.

Miles, Donna. 2005. "National Guard Responds to Hurricane Katrina." *American Forces Press Service* 202.

Northrop Grumman Corporation, News Release. "Northrop Grumman Announces Results of Cash Tender Offer for Gulf Opportunity Zone Industrial Development Revenue Bonds." HYPERLINK "http://investor.northropgrumman.com/phoenix.zhtml?c=112386&p=irol-newsArticle_Print&ID=1501967" http://investor.northropgrumman.com/phoenix.zhtml?c=112386&p=irol-newsArticle_Print&ID=1501967. Accessed August 11, 2016.

Putnam, Robert. 2000. *Bowling Alone: The Collapse and Revival of American Community.* New York: Simon & Schuster.

Reeves, Tate, interview by Marc Landy. 2008. *Mississippi State Treasurer* (January 9).

Ripley, Amanda. 2005. "How the Coast Guard Gets it Right." *Time.com.* http://nwsearleseacadets.org/index/News_files/cgright.pdf. Accessed October 16, 2014.

Robbins, James S. 2005. "Where are the Guardsmen?" *National Review Online.* www.nationalreview.com/article/215311/where-are-guardsmen-james-s-robbins. Accessed October 17, 2014.

Roberts, Patrick S. 2006. "FEMA and the Prospects for Reputation Based Autonomy." *Studies in Political Development* 20: 71.

Sigo, Shelley and HYPERLINK "www.bondbuyer.com/sdm/17.html" Watts, Jim. "GO Zone Cut-Off Looms for Issuers." *The Bond Buyer* September 1, 2010. HYPERLINK "www.bondbuyer.com/issues/119_418/go_zone_cutoff-1016807-1.html" www.bondbuyer.com/issues/119_418/go_zone_cutoff-1016807-1.html. Accessed August 11, 2016.

Sollenberger, Mitchel, and Mark J. Rozell. 2012. *The President's Czars: Undermining Congress and the Constitution.* Lawrence, KS: University Press.

United States Code Title 42. The Public Health And Welfare Chapter 68. Disaster Relief. (n.d.) *Robert T. Stafford Disaster Relief and Emergency Assistance Act, as amended, and Related Authorities.* www.fema.gov/pdf/about/stafford_act.pdf. Accessed September 11, 2014.

U.S. Congress. 2013. "Progress Report: Hurricane Sandy Recovery—One Year Later." http://transportation.house.gov/calendar/eventsingle.aspx?EventID=359869. Accessed November 2, 2014.

U.S. Department of Energy. 2012. "Hurricane Sandy-Nor'Easter Situation Reports." http://energy.gov/articles/hurricane-sandy-noreaster-situation-reports. Accessed December 3, 2012.

U.S. Department of Housing and Urban Development. 2014. " CDBG-DR Eligibility Requirements." *HUD Exchange.* www.hudexchange.info/programs/cdbg-dr/cdbg-dr-eligibility-requirements/. Accessed September 30, 2014.

U.S. Geological Survey. 2012. *Historic Earthquakes.* November 1. http://earthquake.usgs.gov/earthquakes/states/events/1994_01_17.php. Accessed September 23, 2014.

U.S. House of Representatives. 2006. "A Failure of Initiative: Final Report of the Select Bipartisan Committee to Investigate the Preparation for and Response to

Hurricane Katrina." 217. www.gpo.gov/fdsys/pkg/CRPT-109hrpt377/pdf/CRPT-109hrpt377.pdf. Accessed September 10, 2014.

United States Government Accountability Office. 2008. "Gulf Opportunity Zone: States Are Allocating Federal Tax Incentives to Finance Low-Income Housing and a Wide Range of Private Facilities." 8. www.gao.gov/new.items/d08913.pdf. Accessed September 27, 2014.

US Fed News Service. 2006. "Over 5,700 Grant Checks Paid to Gulf Coast Homeowners." *High Beam Research*. November 16. www.highbeam.com/doc/1P3-1163895071.html. Accessed September 27, 2014.

White House. 2006. "The Federal Response to Hurricane Katrina: Lessons Learned." 131. https://georgewbush-whitehouse.archives.gov/reports/katrina-lessons-learned/. Accessed October 8, 2014.

4 Assisting Individuals with Access and Functional Needs

The Intersection of Disabilities, Planning and Disaster Policy

Melissa Pinke, Stacey Mann, and Elizabeth Todak

Introduction

Many Americans with physical or cognitive limitations continue to struggle, despite improvements in educational, structural and social circumstances. Due to various limitations, many people with disabilities require special services or have unique needs related to health care, transportation, housing, employment, education and long-term services. While some governments have made improvements to assist individuals with special needs, many continue to struggle in their normal, day-to-day lives. For some, the slightest change can be detrimental; thus, when disasters strike, many of these individuals may face obstacles too difficult to overcome. Although several of the issues mentioned should be apparent when planning for crises, unfortunately, many details are overlooked or are the responsibility of building owners, managers and individuals.

To better address the issues faced by the most vulnerable of populations, a more relevant descriptor, access and functional needs (AFN), has emerged (Kailes & Enders 2007). Functional-needs planners look beyond disabilities and circumstances to address the needs in relation to emergency planning. People with various disabilities require different accommodations to be able to function in the non-crisis world, thus the same accommodations are needed in times of crisis, especially if the individual must leave home quickly without their necessary equipment. A lack of planning for these needs ensures that individuals who can normally be on their own are forced to depend on others.

Included in the category of AFN are those individuals with mobility and dexterity issues, those requiring medical care (dialysis, respiratory support) and those with limitations due to medical conditions (asthma, arthritis, chemical sensitivities, pregnant women, morbidly obese, mental illness). Also included are minority groups, non-English speaking residents, the homeless population, zero-vehicle households, visitors, those in institutions, children and the elderly (Kailes & Enders 2006; Vogt Sorensen 2006). According to Kailes & Enders (2007), this population represented 49.99 percent of the 2000 Census data, which represents a far larger portion of the community than emergency planners may realize. Therefore, plans that do not address the various needs of all of the individuals in the community are incomplete.

The purpose of this chapter is to investigate the history of functional and special-needs policies and legislation for times of crisis in the U.S.. First, a brief history of special-needs issues in the U.S. and early policies will be discussed. As with many other areas relevant to emergency planning, Hurricane Katrina showed a significant need for planning for individuals who may face great obstacles in times of crisis. The impact of that catastrophe on special-needs planning will exemplify the current issues faced.

Disability History and Policy in the U.S.

People with disabilities have historically faced many struggles. A brief examination of treatment of these individuals shows a dark past of torture and neglect of many types (Pacer Center 2004). However, in the 1700's treatment for children with disabilities emerged, and in Germany the first school for deaf children was founded, with schools in Scotland and England opening soon after. In the 1800's, science replaced religion as the basis of explaining and understanding physical and mental disabilities, which resulted in dispelling myths and increasing acceptance of individuals with limitations (Adams, Bell & Griffin 2007).

Individuals with disabilities have played a significant role in American history from its very beginning. For example, Benjamin Franklin, who at times could not walk due to severe gout, was carried into the Constitutional Congress of 1787, while Stephen Hopkins, who had cerebral palsy, said "My hand trembles but my heart does not" when he signed the Declaration of Independence (Ferleger 2010, p. 27). Some twenty years later, during a 1798 Congressional session, the Act for the Relief of Sick and Disabled Seamen was passed, which called for a small tax on seaman salaries to assist funding for hospitals to address their needs (Jensen 1997). This particular piece of legislation was not only the first healthcare bill, but also the first to attempt to assist individuals with disabilities.

In the early 1800's, schools for the blind began to open (Adams *et al.* 2007), and with approximately 8,000 individuals who had previously been "confined to back rooms and kitchen corners, where they lived out quiet and sedentary lives" in homes of relatives, the introduction of these schools in Boston, Philadelphia, and New York brought a hidden problem into the open (Freeberg 2000, p. 121). During this time, the mentally ill were transferred from jails and poor houses to state mental hospitals, in an attempt to reduce ridicule, physical abuse and neglect (Adams *et al.* 2007), and schools for training those with developmental issues also opened. However, individuals who attended the schools were perceived as dangerous as a result of overcrowding and understaffing, and once again, many individuals with disabilities were not only invisible, but also hidden from society. As explained by Ferleger (2010) "the out of sight, out of mind approach was echoed in society at large in the late 1800s" and laws that kept individuals with visible disabilities out of the public eye were passed by some cities (p. 28).

Unfortunately, although education had opened doors for many at one time, jobs for individuals with disabilities were limited. After the Civil War, disability integration became a real concern, when 30,000 soldiers returning from the war had to readjust to civilian life while suffering from amputations and other injuries (Adams *et al.* 2007). According to Blanck (2008):

> Disabled Union Army (northern) veterans were awarded pensions based on their "incapacity to perform manual labor." This medical model defined disability as an infirmity that precluded equal participation in society and the ability to earn an independent living. As today, in the late 1800s, not all disabilities were regarded equally. Certain stigmatized disorders such as mental disorders and infectious diseases often were deemed "unworthy" of public assistance, and persons with these disabilities faced strong attitudinal discrimination.
> (Para. 2)

In 1935 the Social Security Act was passed to help states provide assistance to a broader group of people, including the elderly and the blind, as well as disabled children and individuals in vocational rehabilitation programs (Adams *et al.* 2007; NCLD 2007). In that same year, the League of Physically Handicapped was formed in New York City to protest disability discrimination (Adams *et al.* 2007). Although these events offered more attention to persons with disabilities, finding jobs still proved to be difficult until World War II, when the supply of the American workforce decreased significantly as men went off to fight the war and suitable replacements were needed. Suddenly, workers who once were considered unemployable, such as individuals with special needs and women, quickly became valuable and "fit to work," and "people with every sort of disability got jobs" (Longmore 2009, p. 11).

Although many lost their jobs upon the conclusion of the war, the years after World War II were marked by legislation that sought to address the rights of individuals with disabilities. In fact, Longmore (2009) states that case law ranging in focus from education, to health care, to employment emerged during these years at both the state and federal levels. "Disability-related legislation and litigation became a major area of governmental and judicial activity" (p. 14), which resulted in the creation of programs and organizations that focused on meeting the needs and rights of special-needs individuals. For example, in 1940, both the National Federation of the Blind and the American Federation of the Physically Handicapped were established. Soon after, in 1943, the Vocational Rehabilitation Amendment was passed, which included physical rehabilitation in vocational programs to allow the participation of people with mental illness and retardation. Five years later, the University of Illinois developed the disabled students' program, one of the first of its kind, which offered services and independent living centers for college students with special needs. Finally, one of the most notable federal programs of the era was a 1950 amendment to Social Security, which gave financial assistance to permanently and totally disabled individuals (Adams *et al.* 2007).

In 1958, accommodations in buildings and other public spaces also came under review when a conference was held by the President's Commission on Employment of the Handicapped, the National Easter Seal Society and the American National Standards Institute to identify standards for making buildings accessible by all. Although the standards were considered voluntary upon their creation, adoption by a state or local government meant they could be enforced, which actually started to occur in the mid-1960s (Welch 1995).

In 1964, the Urban Mass Transportation Act required accessible systems for the elderly and handicapped if federal money was accepted, and in 1968, the Architectural Barriers Act required all buildings built or leased with federal funds to be accessible. Two years later, in 1970, the Urban Mass Transit Act required all new mass transit vehicles to be equipped with wheelchair lifts, although full implementation didn't occur until 1990. In 1973, the Rehabilitation Act initiated the creation of cut-curb sidewalks, designated handicapped parking spaces, ramps in buildings and equal employment practices. Then, in 1975, the Education for All Handicapped Children Act provided funding for school programs for students with disabilities, which led to a 1979 Supreme Court decision that ruled that any programs receiving federal funds must make reasonable accommodations for disabled individuals (Adams *et al.* 2007). The Americans with Disabilities Act became law in 1990, prohibiting discrimination in state and local government, public accommodations, employment, transportation and telecommunications. Even more recently, the Affordable Care Act of 2010 prohibits discrimination on the basis of health status, in order to protect those needing medical care to address disability needs (The Arc n.d.).

Thus, over the past 200 years, many laws continued to improve opportunities to allow those with special needs to live without external restrictions. However, the various pieces of legislation resulted in the emergence of gray areas. The Department of Justice received many questions on the appropriate interpretation of the term "disability" within the Americans with Disabilities Act of 2008 so as to determine if discrimination was evident. The definition of disability has been revised, but according to the Americans with Disabilities Act, amended in 2009, a disability is "a physical or mental impairment that substantially limits one or more major life activities of such individual; a record of such an impairment; or being regarded as having a disability" (Morris 2011, p. 6). Although the Act does not define or discuss the diagnoses that would lead to categorization as a disability, the specific conditions that meet the criteria of a disability were determined, and include: deafness, blindness, intellectual deficiencies, partial or completely missing limbs, mobility impairments, autism, cancer, cerebral palsy, diabetes, epilepsy, HIV infection, multiple sclerosis, muscular dystrophy, major depressive disorder, bipolar disorder, post-traumatic stress disorder, obsessive-compulsive disorder and schizophrenia (ADA National Network FAQ).

Special Needs, Emergency Management and the Law

Since government began recognizing that some populations needed assistance both during times of normalcy and times of crisis, the terms referring to those most vulnerable have evolved. Understanding the differences used in AFN literature will provide some insight into this population's lack of inclusion in disaster planning. For example, the ADA offers the following:

> An individual with a disability is defined by the ADA as a person who has a physical or mental impairment that substantially limits one or more major life activities, a person who has a history or record of such an impairment, or a person who is perceived by others as having such an impairment.
>
> (Morris 2011, p. 6)

This term simply defines disability, but does not state that a person is incapable of activity. With the advancement of medicine, many life activities are now accomplished with assistance. Another frequently used term within emergency management is "special needs," which is defined by the state of Florida as "someone who, during periods of evacuation or emergency, requires sheltering assistance, due to physical impairment, mental impairment, cognitive impairment, or sensory disabilities" (Special Needs Interagency Committee 2012, p. C2). However, as noted by Kailes and Enders (2007), "it is important to think broadly about disability in terms of function and not in terms of impairment of diagnosis" (p. 233). A broad definition of access and functional needs (AFN) is also included in FEMA's shelter guidance document, which states:

Those actions, services, accommodations, and programmatic, architectural, and communication modifications that a covered entity must undertake or provide to afford individuals with disabilities a full and equal opportunity to use and enjoy programs, services, activities, goods, facilities, privileges, advantages, and accommodations in the most integrated setting, in light of the exigent circumstances of the emergency and the legal obligation to undertake advance planning and prepare to meet the disability-related needs of individuals who have disabilities as defined by the ADA Amendments Act of 2008, P.L. 110-325, and those associated with them. Access and functional needs may include modifications to programs, policies, procedures, architecture, equipment, services, supplies, and communication methods.

(p. 41)

Although the federal government has adopted policies regarding assistance to AFN individuals during times of normalcy, some debate about the extent of assistance, or the responsibilities of government at all levels, does exist. Thus, when disasters such as Hurricane Katrina, Hurricane Sandy or the tornado outbreak of April 2011 occur, problems are quickly made evident, both by politicians and public administrators and the public.

During times of crisis especially, but also during times of normalcy, emergency management faces the obstacle of limited resources. For example, within local governments politicians focus on re-election, so spending on roads and schools, which are seen as immediate needs, are much more likely to gain public approval than mitigation and preparedness projects that plan for a disaster that may or may not occur (Mann 2014; Donahue and Joyce 2001). In addition, budget cuts present major obstacles. At the local-government level, decreased budget allocations mean that emergency-management agencies have to do more with less, both in terms of actual spending and with staff. Budget shortages result in layoffs or freezes on hiring, so fewer personnel are available to participate in emergency planning. As mentioned above, planning for the AFN population within a community requires a significant amount of time. "At a minimum, planners must identify at-risk individuals through registries, delegate authority and responsibility to appropriate governmental officials, collect supplies, and allocate resources, among other steps" (Hoffman 2009, p. 1496). Without adequate personnel, not all projects will be priorities, and for vulnerable populations this could be detrimental, resulting in the inability to evacuate their homes, the inability to obtain medical supplies at distribution points or even the inability to understand emergency messages (Hoffman 2009). Planning for AFN individuals means ensuring that shelters have text telephones, open-caption television and individuals to assist, such as American Sign Language interpreters. Inappropriate time and attention to AFN planning could result in the needs of some members of the community being overlooked, and a less prepared population may result in increased demand for precious limited resources. Kailes and Enders (2007) argue that "as the term is typically used, the special-needs population makes up at least half of the U.S. population" (p. 230), which is too broad and does not specify true groups of individuals with needs, thus impacting proper planning and response. In addition, they add that "it is important to think broadly about disability in terms of function and not in terms of an impairment or diagnosis" (p. 233).

Emergency managers also must consider immobility, a very specific AFN group, which would be an important life-safety factor in the mission of disaster response. Fox, White, Rooney, and Rowland (2007) surveyed emergency-management agencies that had experienced disasters from 1998 to 2003 and examined whether or not the disasters affected future planning for special-needs individuals. The focus of the research, mobility, was analyzed by reviewing plans both pre- and post-disaster, as well as by surveying emergency-management directors. Four areas were examined: county programs, practices and policies, assessment of risk and assurance and policy development (Fox *et al.* 2007). The three recommendations from the study for emergency managers to consider were better inclusion of the AFN community in planning, better training regarding special-needs response and identification of AFN individuals in the community.

However, some may argue that, especially when limited, the resources available for emergency-preparedness should be allocated in a way that does the most good for the most people. Focusing on a small segment of the community, some might say, does not lead to the maximum benefit for the community as a whole. On the other hand, others argue that because those most vulnerable during times of normalcy become even more at risk when disaster strikes (Finch, Emrich, & Cutter 2010), and because disasters "do not affect all members of society equally", vulnerable populations have the greatest amount to lose, so more resources should be allocated to helping them (Fothergill & Peek 2004, p. 89). Further, Fothergill and Peek (2004) argue that "poor people around the world suffer the greatest disaster losses" (p. 91), while Hoffman (2009) adds that "Many commentators argue vigorously that considerable resources should be invested in assisting vulnerable individuals such as the poor, institutionalized, and disabled, because they are the least able to withstand the hardships caused by catastrophic events" (p. 1513).

Thus, the ethical questions that arise as a result of a disaster initiate discussions that often are answered by current laws, or may even result in the creation of a new policy. Schwab and Beatley (2013) present several arguments regarding emergency-management based on legal perspectives. In the area of constitutional law, the right to equal protection ensures that all races are protected from discrimination. "If authorities deliberately underserve or mistreat a minority community, such as African Americans or other minority group during an emergency because of race, that group could have a valid equal protection claim" (Schwab & Beatley 2013, p. 59). Further, the authors state that because of legislation such as the Americans with Disabilities Act and the Civil Rights Act of 1964, individuals are protected from discrimination on the basis of disabilities, age, race, color or national origin. Disasters often result in locales receiving financial assistance as a result of a presidential disaster declaration, thus individuals who are affected cannot be excluded from any program that offers public assistance for any one of the protected class characteristics.

At times, disasters present new problems or ethical issues that are not addressed by case law or legislation. When the problem receives significant attention by the public, policymakers may engage in the learning process or "the process by which participants use information and knowledge to develop, test, and refine their beliefs" (Birkland 2006, p. 8). The attempt to learn about a particular problem is sometimes a result of a "focusing event," which is defined by Birkland (2006) as

> an event that is sudden, relatively rare, can be reasonably defined as harmful or revealing the possibility of greater potential future harms, inflicts harms or suggests potential harms that are or could be concentrated on a definable geographical or community of interest, and that is known to policymakers and the public virtually simultaneously.
>
> (p. 2)

These types of events often highlight that some sort of policy failure has occurred and that changes based on the lessons learned from the event should be considered, discussed, designed and implemented. Unfortunately, the U.S. became all too familiar with these issues in 2005.

Special Needs Issues during Hurricane Katrina

No other event in U.S. history highlighted the lack of planning and preparedness for vulnerable populations on such an extreme scale as Hurricane Katrina in 2005. Many criticized New Orleans' preparedness, especially after its conducting an eight-day hurricane exercise, with more than fifty local, state, federal and volunteer organizations and over 250 officials, a year prior to Katrina (Beriwal 2006). The second phase of the exercise, which was conducted in July 2005, included transportation issues. However, many critics of the response do not realize that the second exercise in the Hurricane Pam study was conducted just one month prior to Hurricane Katrina's landfall, which did not offer enough time or opportunity to implement many changes. Although the group offered generalized findings, a specific category dedicated to the AFN population was not created (Beriwal 2006). The Hurricane Pam exercise was the first of its kind, so a month or even a year is not likely to be a realistic timeframe for any municipality to develop a plan that could be implemented so quickly. Unfortunately, the horrific devastation that occurred in New Orleans and surrounding areas shone a spotlight on the importance of planning for the individuals who need it the most, and no area was more evident than transportation needs. Whereas issues with medical and special-needs shelters were covered during the second exercise, transportation challenges, especially with regard to transporting AFN individuals, were neither identified nor addressed, which presented major issues when evacuation was deemed mandatory.

On Friday August 26, 2005, the National Hurricane Center predicted landfall of Hurricane Katrina in Plaquemines Parish, which is located in southeast Louisiana, and the storm was expected to hit as a Category 4 hurricane (Schleifstein & Krupa 2008). Many Louisiana residents got a clearer picture of the hurricane when it was described by Jefferson Parish Emergency Management Director Walter Maestri as "following Hurricane Betsy's track … with the strength of Hurricane Camille" (Nolan 2005, p. 1). Although Governor Blanco declared a state of emergency on Friday, Nagin issued only a voluntary evacuation on Saturday at 5 p.m., less than 48 hours before Katrina's landfall. The *Times-Picayune* reported:

> Nagin said late Saturday that he's having his legal staff look into whether he can order a mandatory evacuation of the city, a step he's been hesitant to do because of potential liability on the part of the city for closing hotels and other businesses.
>
> (Nolan 2005, p. 1)

It wasn't until 9:30 a.m. on Sunday, August 28 that Nagin ordered a mandatory evacuation (Katrina Timeline 2008). In an interview with National Public Radio, Nagin took responsibility for the late evacuation, but stated the city could be liable for shutting down businesses and hotels. However, because decision-making in times of disaster requires that "long-standing policies" be put aside, Nagin was held responsible for not attempting to evacuate those left behind. Although the mayor argued that the city did not have enough buses to evacuate the 100,000 people left behind, adequate transportation was available for New Orleans residents to be shuttled to the Superdome (Martinko *et al.* 2009).

On Thursday, August 25, the Louisiana Office of Homeland Security and Emergency Preparedness (LOHSEP) began issuing warnings to parish and state agencies, while on Friday, August 26, the Louisiana Department of Transportation and Development (DOTD) began planning the coordination of evacuation for individuals with special needs. In addition, the Louisiana Hurricane Task Force held a conference call the Friday afternoon prior to Katrina's landfall and the state police announced its plans to open the Emergency Operations Center (EOC) in Baton Rouge on Saturday, August 27 (U.S. Senate 2006). In the meantime, Blanco began working with each of the agencies, which allowed "the State of Louisiana [to activate] more than twice the number of National Guard troops called to duty in any prior hurricane, and achieved the largest evacuation of a threatened population ever to occur" (United States Senate 2006, p. 5). However, although Blanco called for more state troops before Hurricane Katrina hit, assistance from the federal government was not requested until the day after landfall, which meant additional troops did not arrive for several days.

The fact that the City of New Orleans and the State of Louisiana would not be able to accommodate all of the residents without transportation and supplies was noted before the storm's landfall, but not addressed. In their report, *Hurricane Katrina: A Nation Still Unprepared* (2006), the U.S. Senate claimed:

> Mayor Nagin and Governor Blanco—who knew the limitations of their resources to address a catastrophe—did not specify those needs adequately to the federal government before landfall. For example, while Governor Blanco stated in a letter to President Bush two days before landfall that she anticipated the resources of the state would be overwhelmed, she made no specific request for assistance in evacuating the known tens of thousands of people without means of transportation.
>
> (p. 6)

In fact, Blanco did not request that the federal government send additional buses until Monday, the same day Katrina made landfall; the buses did not arrive until Wednesday, although it was known prior to Katrina's arrival that not everyone would be able to evacuate on their own. Rather than addressing the issue prior to the hurricane, the governor did not begin coordinating an effort to locate buses and to commandeer school buses until Tuesday, and it was not officially approved until Wednesday (U.S. Senate 2006). In addition, Blanco overlooked the fact that Orleans Parish would need financial assistance after Katrina made landfall when, in a letter to President Bush, she requested $9 million using the Federal Stafford Act. She later made the correction when she sent a second letter on Sunday requesting $130 million in aid, but still neglected to ask for evacuation support or transportation (U.S. Senate 2006).

Obviously, while some proactive decisions were made, not all planning and decision-making was effective. In fact, city and state officials were well aware of the vulnerable position New Orleans and the rest of the state were in if a major hurricane hit, but neither appropriated enough funds or resources to emergency planning. According to the National Response Plan (NRP), ensuring that emergency alert systems are capable of full functioning is not the responsibility of local mayors, FEMA officials or the president. The responsibility lies with governors. In fact, the NRP states, "As a State's Chief Executive, the Governor is responsible for the safety and welfare of the people of that State ... [and] for coordinating State resources to address the full spectrum of actions ... to prepare for [and] respond to ... natural disasters" (U.S. Senate 2006, p. 149).

Unfortunately, challenges faced during Katrina are difficulties faced daily by special-needs populations. These include issues with communications, transportation, accessibility, housing, employment, education and healthcare access. For example, those with hearing difficulties do not always have access to closed captioning, sign language or any written explanation of charts and graphs on news channels. When television is not an option, the hearing-impaired cannot simply turn on a radio to obtain updates. Further, written instructions and other visuals are often unrecognizable or difficult to interpret to those with vision limitations (Frieden 2006). If alternate communication methods are not fully functional on a daily basis, they are unlikely to be working and available during emergencies. If access to transportation is difficult or costly for routine matters, it should be expected to get even worse when these resources are in high demand. Therefore, vulnerable residents and emergency-management organizations need to do further planning.

During the response to Hurricane Katrina, many issues arose around leadership and organizational failures, inadequate regulations and overwhelmed systems (Bea *et al.* 2006), and in the last days of August 2008, residents of New Orleans and surrounding areas watched as Hurricane Gustav made a steady approach toward the coast. With Katrina still fresh on their minds, many wondered if the lessons government officials claimed to have learned would actually be put into practice effectively. Editorials and news stories in *The Times-Picayune* were published while Gustav was still 155 miles southeast of Cuba, predicted to strengthen to a Category 3 storm while just 300 miles south of the mouth of the Mississippi. New Orleans was in the "cone of error" (Schleifstein & Krupa 2008). Although Hurricane Gustav did not make landfall until Monday, September 1 as a Category 2 hurricane, 80 miles southwest of New Orleans in Cocodrie, Louisiana, early preparations, quicker decision-making and more cooperative relationships were evident in the days leading up to the storm.

In the months and years after Hurricane Katrina, the City of New Orleans, the State of Louisiana and departments of the federal government began to make preparations for future hurricanes in New Orleans. While many issues came out of Hurricane Katrina, the most evident ones pertained to evacuation, slow response and uncertainty over authority. At the city level, the initial evaluation is that officials did learn lessons regarding evacuation. First, discussion of evacuating the city began five or more days before the hurricane made landfall. Both Mayor Nagin and the city's director of Homeland Security and Emergency Preparedness, Jerry Sneed, did not seem to hesitate to discuss evacuation, with promises of follow-through.

The evacuation plan that was created came with several steps and a definite timetable. Not only did the new evacuation plan include details for city evacuees; also it synced with a regional plan, directing residents from varying locations to different evacuation points. The hope was that congestion during contraflow, having traffic move in only one direction, would be reduced (Schleifstein & Krupa 2008). In addition, in the days of Katrina, much criticism centered on the fact that Nagin knew that about 100,000 residents, including the sick and elderly, would not be able to evacuate, so he opened the Superdome as a shelter. Learning the lessons of Katrina, a new plan, the City Assisted Evacuation Plan, offered those who could not evacuate on their own an opportunity to call 311, register with the city and then be provided transportation to a shelter out of town. Residents were told to go to one of seventeen sites in the city where buses would transport them to the Union Passenger Terminal, to then board a train and be chartered by Amtrak to cities in Louisiana and Mississippi. Those who could not get to one of the sites could call for assistance and be picked up at their home. Although in the week before Gustav—when Tropical Storm Fay threatened New Orleans—only 7,000 of the 30,000 estimated to need assistance had registered with the service, the Gustav evacuation was labeled "the largest government-assisted evacuation in New Orleans' history" (A Wet Run 2008, p. 6; Carr 2008, p. 1). Although some residents had to wait in long lines for registration and some shuttles for handicapped and sick residents did not show, approximately 18,000 people fled the city on government-provided buses, trains and planes (Carr 2008). Some did criticize the evacuation effort, stating that evacuees were not told of their destination and then faced unsanitary conditions once they arrived at one of four state-run shelters (*New York Times* 2008). Public officials quickly acknowledged that the evacuation was not perfect, but that the disaster plans had undergone a complete make-over following Katrina and now just needed "finessing" (Carr 2008). John Renne, an assistant professor at the University of New Orleans who studies evacuation issues, said, "I don't think there was anyone who didn't realize they had the ability to evacuate" (Carr 2008, p. 9). It was also made clear that those who chose to stay during the storm could not expect local first responders to help (Pope & Krupa 2008).

Contrary to its stance on Katrina, the State of Louisiana became involved with the preparations for Gustav and took a proactive stance in decision-making several days before Gustav entered the Gulf of Mexico.

Governor Bobby Jindal, elected on October 20, 2007, said a state of emergency could be declared as early as four days before Gustav made landfall, which would assist in the evacuation process. Contraflow could begin more than forty-eight hours prior. In addition, the state organized shelters and "identified 10,000 critical care beds and 68,000 regular beds for evacuation" (Schleifstein and Krupa 2008, p. 1).

The revised state evacuation plan is initiated eighty-two hours before tropical storm force winds hit the coast of Louisiana, which, in the case of Gustav, actually began on August 30. When the company with which the state had contracted to provide 700 buses was unable to fulfill the commitment, the governor quickly made the decision to request buses from the RTA (Barrow & Scott 2008, p. 1). Jindal was praised in the media for his quick decision-making and take-charge attitude. Douglas Brinkley, author of the book *The Great Deluge*, said, "He promised to be a hands-on administrator, and I think he delivered. He had such an easy factual grasp of the situation. It's almost the exact opposite of Blanco and Nagin during Katrina" (Whoriskey 2008, p. A06). Whereas it was agreed that Gustav was no Katrina, Jindal was praised for his readiness and decisiveness that covered areas from healthcare to evacuation.

Special Needs Registries

One way in which these needs have been addressed is through the creation of special-needs registries—a way to collect information regarding special-needs community members and identify those needs in order to provide assistance (Black 2010; Hewitt 2013). By identifying the different types and quantities of needs, emergency planners can coordinate assistance. Florida was the first state to establish a special-needs registry (The Florida Legislature 2015). Title XVII of the Florida Statutes § 252.355 established the requirement for a registry in each county, comprising special-needs individuals that will require assistance with evacuations or sheltering. This law went into effect on March 1, 2015, and specifically requires the registration of any person that is a client with a disability who is in a state or federally funded service program (The Florida Legislature 2015). Healthcare services, related support agencies and pharmacies are expected to provide information to those who may need assistance, and are also able to register individuals directly in the database. Utilities must also provide information to customers upon initial sign-up and at least twice between January and May (The Florida Legislature 2015). This program went live in April 2015, and the Florida Special Needs Working Group conducted an initial evaluation of the program on August 8, 2015; minor changes have been made and additions considered (Stoughton 2015).

As with any new program, start-up can be slow. At their evaluation, the working group reported that sixteen out of sixty-seven counties in Florida have provided registry data; of these sixteen, only eleven counties have more than forty registrants (Stoughton 2015). Each county handles the registry in various ways. For example, Pasco and Miami-Dade Counties require the registration form to be printed, completed and mailed (Pasco County, Miami-Dade County, 2015). Osceola requires a phone-in registration (Osceola County 2015). Orange County lists multiple ways for participants to register (Orange County 2015).

Florida is not the only state to provide special-needs registries. In a review by the authors of all fifty states' emergency-management websites, thirty-six have registries in at least one county. Eleven have no easily identifiable formal registries, but do provide special-needs information. One state, South Dakota, provided no additional information or registry information that was easily found for persons with special needs.

The Post-Katrina Emergency Management Reform Act of 2006 provided a way to focus FEMA on a new mission to prepare for, rather than waiting for, requests for help, especially with regard to assisting AFN individuals (Department of Homeland Security 2015). This proactive approach was one of the biggest changes since 9/11. Some of the key changes included appointing a disability coordinator, requiring a family registry to locate separated family members, assisting with evacuation needs and addressing unmet needs through case management (Emergency Management Institute n.d.).

The Post-Katrina Act addresses assistance and accommodations for both persons with disabilities and those with limited English proficiency. Information must be made available to everyone in a language they understand, while accommodations and durable medical equipment are required to be available for those that need it. Service animals as well as pets must also be included in assistance programs (Bea *et al.* 2006). An important point is that vulnerable populations go beyond those with disabilities and limited English, including those that need other sources of assistance as well.

While Hurricane Katrina highlighted the challenges of the disabled population, it is important to note that most of the population was able to evacuate as planned (Black 2010). Many lawsuits have since emerged around the lack of planning for special-needs populations. Some worth mentioning include the City of Los Angeles (Disability Rights Legal Center 2011), City of Oakland, CA (Disability Rights Advocates 2010), the District of Columbia (Disability Rights Advocates 2014), New York City (Santora & Weiser 2013) and a current lawsuit in Washington, DC (Beck 2014). Each of these cities will have to take a realistic look at their emergency plans with respect to special-needs populations and make changes accordingly.

Thus, one solution may be for emergency managers to include information on vulnerable populations in the normally distributed emergency-preparedness materials or regular presentations to the public, which would allow all members of the community an opportunity to learn about at-risk individuals. The result, then, may be that citizen involvement in offering assistance would increase, thus enabling emergency managers the opportunity to use resources on other projects.

While those in institutions will require the most advanced care during emergencies, the respective organizations are required to have plans in place to support their own residents, so initial response should not be the direct responsibility of the state or county. It would be prudent, however, to meet with these facilities and understand the extent of their planning. If their plans are non-existent, incomplete or inadequate, county and state resources will be requested during an emergency. Meetings to discuss institutional plans will also help identify the supporting resources that are already committed, how they can be shared or if more are needed.

Current Issues

Several current issues surround the use of emergency-management registries. While the use of special-needs registries sounds like a good approach, there can be limitations. A study conducted in 2007 examined specific issues and challenges in the use of registries by individuals with limiting abilities or without transportation. Key points raised by the respondents were related to identifying those in need, resistance of undocumented immigrants due to fear of deportation, logistics around resources related to evacuations and the willingness of persons to evacuate. An important factor is that some people are embarrassed or have concerns regarding invasion of privacy and do not want to disclose disabilities. Other individuals are busy and forget to go in and register, while some have limitations that make filling out paperwork or online forms too problematic. Some people think they are safer in their homes and would rather face the risk than leave or go to a shelter. Those with pets would rather stay with their pets, while low-income individuals are concerned with expenses. For some with medical needs, leaving behind specialized equipment is a major concern (Black 2010). By addressing each of these issues, improvements can be made.

Hewitt (2013) identified another challenge which registries may have created. When someone with a disability reports their needs to an authority, a contract is established, and that authority is obligated to ensure the person's needs are met during an emergency. This can cause problems on two levels. First, there may be an expectation by the individual that help will come; second, if help does not come, liability issues arise (Hewitt 2013). Thus, clear communication regarding the meaning of enrolling in a registry is necessary (Black 2010).

Finally, specific standards on how to collect or manage data have been established (Hewitt 2013), and the programs are still too new for best practices to have emerged. Emergency managers are reluctant to use registries because of cost, perceived ineffectiveness and low participation, as well as lack of funding and resources to maintain and support both the registry process and the needs of individuals (Hewitt 2013). Hewitt also noted that using support agencies to assist with enrolling AFN individuals into registries only yielded 4 per cent of the registries' participants, with one reason being confidentiality issues and the second being the additional work that comes with doing so. While working with agencies to build a registry is a good idea, it would likely require communication with these agencies to clarify misunderstandings and HIPPA requirements (U.S. Department of Health and Human Services 2014).

The Federal Highway Administration also has guidance on evacuating special-needs populations. In its document *Evacuating Populations with Special Needs*, it states that "these are untested ideas that may carry as many risks and challenges as do potential benefits" (Houston, Vann Easton, Davis, Minicin, Phillips & Leckner, 2009, p. 20). These issues will need to be addressed.

Many previous assumptions regarding assistance for AFN individuals are no longer acceptable. According to ADA National Network, (2014) an individual with a disability cannot be required to bring an interpreter or a buddy to obtain information, unless:

> It is an emergency involving imminent threat to the safety or welfare of an individual or the public where there is no interpreter available; or when the individual with a disability specifically requests that the accompanying adult interpret or facilitate communication, the accompanying adult agrees to provide such assistance; and relying on that adult is appropriate under the circumstances.
> (Sign Language Interpretors, para. 1)

However, with all of the focus on individuals' inclusion into emergency-preparedness planning, these exceptions should no longer be part of any plans. Individuals should get information directly. Proactive planning for those needing assistance is just as crucial in the workplace, where there are 33 million people, or 18.6 percent of working adults, with some sort of disability, per the 2000 Census. Emergency plans and local requirements—especially in NYC—assisting those with disabilities have dodged specific planning methods, relying on the use of a buddy system; however, even FEMA agrees problems can come along with this option (Polanco 2004; FEMA n.d.). Generally, people want to be as independent as possible, and rely on their own means to obtain information rather than rely on others. Due to absences, meetings and overall day-to-day activities, the buddy system can be ineffective and, without practice, forgotten. Keep in mind that in the workplace, employees manage to get ready for work, arrange for transportation, arrive at their workstations and conduct business for the company; these employees include those with disabilities. Therefore people with disabilities have many abilities to cope with daily life activities. With careful thought, the assistance needed for emergency response measures can and should be included to address self-sufficiency.

Assisting with the needs of those with disabilities should be included in building location emergency action plans. It is necessary to ensure everyone is notified of emergencies, addressing the needs of those with both hearing and vision impairments. Occupants must be able to understand directions, so use simple phrases and the languages of those likely to be present. The third biggest area to address is egress, or getting out of a building or to a safe area. Consideration must be in place for situations where elevators cannot be used and those with mobility impairments or health conditions may have difficulty using stairs to evacuate. In-house emergency teams are required to be trained by OSHA; covering specific details related to assisting those with special needs in facilities should be sufficient. This may include having those with special needs on floors above or below ground level report to a specific, safe location to obtain additional assistance. If someone is likely to have a special need but doesn't disclose it, plan for it anyway. The goal is to focus not on the disability, but on the type of assistance needed. The reality is that other employees or visitors may have similar needs, so including these provisions will improve emergency plans. People with assistance needs are not separate from the community, but are within the community. In order to be effective, planning needs to be inclusive of all groups and not considered a separate program (Kailes & Enders 2007).

As an example of federal guidance, the Federal Highway Administration recommends starting response measures earlier for those that need assistance. If evacuation is required, those with special needs must be the first to go in order to avoid transportation delays, which can compromise care and increase health risks and possibly even lead to death. Emergency management, transportation agencies and those with special needs should have open dialogue and create a planning group to identify requirements, strategies and timelines, while at the same time building trust (Houston *et al.* 2009). It is important to note that buses or vans may have limited capacity for wheelchairs and scooters, likely increasing the number of transportation assets required. If these buses and vans also are contracted for deployment to assist other nursing homes, hospitals or other facilities, other transportation companies will need to be contacted ahead of time. Even shelter-in-place measures can take longer for people with some disabilities.

Suggestions for Registries

Based on this review, several suggestions can improve practices regarding emergency-assistance registries. First, individual needs can generally be grouped into five distinct categories: communications, maintaining independence, supervision, transportation and medical care (Kailes & Enders 2007). These groups include those who have problems with vision, hearing, mobility, cognitive abilities and medical conditions, as well as those with temporary disabilities related to recent illness or injury and those with functional needs (Vogt Sorensen 2006). Depending on the composition of the community, for planning purposes, assume 20–50 percent of the population will need this kind of assistance, regardless of whether the registry numbers substantiate this (Kailes & Enders 2007).

Official communications must provide information in ways that people can both understand and receive, in order for them to act. Information should be provided using simple terminology. Written information should be in large print, and provided in the various languages used in the community. An audio format is also recommended. Information dissemination will be improved by establishing specific times and places to post information and provide interpreters that are trusted by the audience (Kailes & Enders 2007).

Those who need assistance with maintaining independence may require limited medical support, replacing required medications or lost adaptive devices. These individuals may require assistance with activities of daily living, such as feeding, dressing, grooming, bathing and toileting (Kailes & Enders 2007). Pharmacies and home healthcare agencies can be valuable resources to assist with these needs.

Supervision care may be necessary for those with dementia, psychiatric conditions or Alzheimer's. This also includes oversight for unaccompanied children or those with anxiety about strange places and routines (Kailes & Enders 2007). Again, home-care agencies and social-service groups may be able to provide guidance and means to address these resources.

Transportation needs can vary considerably. The most important consideration is how to get people away from unsafe areas to either further transportation hubs or shelters. Protocols stating who will do what and when must be in place. Depending on the needs in the community, specialized transportation may be needed for wheelchairs and motorized devices, which may require many specialized vans. Transportation addresses not only vehicles, but also a method to keep traffic moving on the roads. Readjusting traffic signals to accommodate larger traffic volume should be considered (Wolshon 2009). It is important to test the call-up process often to ensure that the correct number of buses and drivers can respond, and that they are able to access the correct pick-up locations (Houston *et al.* 2009).

Medical support includes resources for patients with conditions requiring observation or medical treatment. This will require more intensive resources, such as medication management, IV administration, dialysis, tube feeding, suctioning, oxygen use, respiratory treatments, caring for catheters, ostomies, wounds and the use of power-dependent medical equipment (Kailes & Enders 2007). Again home healthcare agencies, medical providers and medical-equipment companies are the best resources for this type of care.

Regardless, the safety of vulnerable populations begins with education. Houston *et al.* (2009) recommend that:

> Persons with disabilities engage in self-education, personal planning and preparedness. People with disabilities should develop a personal support network and an evacuation kit. People with disabilities will be motivated to evacuate when they believe that those assisting them are truly ready to meet their individual needs. Furthermore, building trust between those at risk and those involved in evacuation is important. Individuals, particularly the elderly and those with serious health problems, are most likely to evacuate when they trust in the credibility of local officials.
>
> (pp. 21–22)

Individuals who may need assistance must engage in the planning process if they want to be a part of the decision-making process, otherwise policy-level and operational decisions will be made without their input. These groups will benefit from education outreach and materials designed for easy interpretation (Houston *et al.* 2009). Having a detailed plan in place before an event can improve self-sufficiency, requiring fewer resources when an event occurs. Emergency planners can prepare people by providing educational sessions, pamphlets, internet information and individual discussions. This level of need provides an excellent opportunity for resources such as Community Emergency Response Teams (CERT) to focus on educating neighbors and helping others make a plan. Often, a neighbor will be more trusted than personnel from a government agency. This level of interaction encourages neighbors to help neighbors, providing insight into the hazards, resources and solutions available at the community level.

Conclusion

This chapter has attempted to review some of the challenges of individuals with disabilities over the past 300 years and how focusing events have affected policies attempting to assist AFN individuals, both during times of normalcy and times of crisis. The recent development of special-needs registries has made many people anxious about their continued development, uses and limitations, prompting a review of some of the current systems and concerns. However, after witnessing the catastrophe of policy failures during and after Hurricane Katrina, policies regarding how to assist these individuals during the times at which it is most needed clearly must be established.

The U.S. has a long history of assisting individuals with special needs, but to correct leadership failures, new legislation and programs have been developed, discussed and implemented to ensure previous failures do not occur again. One of these programs is the use of special-needs registries. Every community must know their respective populations and understand their needs, as well as find ways to address them. While careful planning should identify how many individuals have special needs, we should also include a certain percentage for those that do not report for fear of stigma, through ignorance of the process or due to overlooking it. With increases in travel, there will likely be a transient group that will be unknown, but must still be planned for. Once the needs of the population are determined, identify how they will be assisted and the resources needed. Ensure the purpose of your registry is clear; explain what can be expected and what information is needed. Keep in mind, the more complicated the process, the greater the chances people will not register. It is always best to encourage self-assistance, but realize this may not be practical for everyone. The success of a community's emergency-preparedness will be measured on how many people are left behind. It is much better to assist those in need before a disaster than to increase rescue efforts during or immediately after an event. By involving the access and functional-needs community and their representatives in the planning process, we can reduce wrong assumptions and focus on legitimate needs, reduce costs and get necessary provisions in place while planning time is available, rather than taxing resources during a disaster. Whereas lessons will always be learned during disasters, learning lessons from previous mistakes will ensure better protection for the whole community.

Bibliography

Adams, Maurianne, Lee Anne Bell, and Pat Griffin. 2007. *Teaching for Diversity and Social Justice*, Second Edition. London: Routledge.
ADA National Network. n.d. "Frequently Asked Questions: What Is the Definition of a Disability under the ADA?" Retrieved May 25, 2016 from: https://adata.org/faq-page#t7n1270
ADA National Network. 2014. "Effective Communication." Retrieved 5 December, 2015 from: https://adata.org/factsheet/communication
The Arc. n.d. "Civil Rights Issues for People with Disabilities." *Public Policy and Legal Advocacy*. Retrieved February 26, 2016 from: www.thearc.org/what-we-do/public-policy/policy-issues/civil-rights

Barrow, Bill, and Robert Travis Scott. 2008. "State Scrambles to Find Buses for Evacuation; Original Contractor Reneged, Jindal Says." August 30. *Times-Picayune*, p. 1.

Bea, Keither, Elaine Halchin, Henry Hogue, Frederick Kaiser, Natalie Love, Shawn Reese, and Barabara Schwemie. 2006. "Federal Emergency Management Policy Changes after Hurricane Katrina: A Summary of Statutory Provisions." *Congressional Research Service*, The Library of Congress. Retrieved August 4, 2016 from: www.fas.org/sgp/crs/homesec/RL33729.pdf

Beck, N. 2014. Disability-Rights Advocates Sue over D.C.'s Emergency Plans. Legal Times. Retrieved 5 December, 2015 from: www.dralegal.org/pressroom/press-releases/district-of-columbia-sued-for-failure-to-serve-people-with-disabilities

Beriwal, Madhu. 2006. "Preparing for a Catastrophe: The Hurricane Pam Exercise. Statement before the Senate Homeland Security and Governmental Affairs Committee, January 24." IEM. Retrieved 5 December, 2015 from: www.hsgac.senate.gov/download/012406beriwal

Birkland, Thomas A. 2006. *Lessons of Disaster: Policy Change After Catastrophic Events*. Washington, D.C.: Georgetown University Press.

Black, Laura A. 2010. Evacuation of Carless Populations: Emergency Management Registries: Merging of Research into Practice. University of Delaware, 2008. Retrieved 5 December, 2015 from: www.ce.udel.edu/UTC/Presentation2010/Laura%20Black%20Evacuation%20of%20Carless%20Populations.pdf

Blanck, Peter. 2008. "Employment of Persons with Disabilities: Past, Present, and Future." *American Psychological Association Spotlight on Disability Newsletter*. Retrieved 5 December, 2015 from: www.apa.org/pi/disability/resources/publications/newsletter/2008/08/employment.aspx

Carr, Sarah. 2008. "Evacuation is Mix of Success, Distress; Officials Boast, but Some Seek Changes." *Times-Picayune*, September 7. p. 9.

Department of Homeland Security. 2015. "Disasters Overview." Retrieved 5 December, 2015 from: www.dhs.gov/disasters-overview

Disability Rights Advocates. 2014. "District of Columbia Sued for Failure to Serve People with Disabilities During Disasters." Retrieved 5 December, 2015 from: www.dralegal.org/pressroom/press-releases/district-of-columbia-sued-for-failure-to-serve-people-with-disabilities

Disability Rights Advocates. 2010. "Sweeping Settlement Reached by the City of Oakland to Include People with Disabilities in Disaster Planning," 2010. Retrieved 5 December, 2015 from: www.dralegal.org/sweeping-settlement-reached-by-the-city-of-oakland-to-include-people-with-disabilities-in-disaster

Disability Rights Legal Center. 2011. "Individuals with Autism Spectrum Disorder and Emergency Planning Services". Minnesota's Governor's Council on Developmental Disabilities. February 11, 2001. Retrieved 5 December, 2015 from: http://mn.gov/mnddc/emergency-planning/la-ep-lawsuit.html

Donahue, Amy K., and Phillip G. Joyce. 2001. "A Framework for Analyzing Emergency Management with an Application to Federal Budgeting." *Public Administration Review* 61, 728–740.

Emergency Management Institute, Federal Emergency Management Agency. n.d. "Lesson 1: Emergency Management Overview." Retrieved May 25, 2016 from: https://emilms.fema.gov/IS230c/FEM0101summary.htm

FEMA. 2010. "Guidance on Planning for Integration of Functional Needs Support Services in General Population Shelters." BCSF Health & Human Services. Retrieved 5 December, 2015 from: www.fema.gov/pdf/about/odic/fnss_guidance.pdf

FEMA Independent Study. 2014. "Emergency Management." Retrieved 5 December, 2015 from:https://training.fema.gov/is/crslist.aspx?page=1.

FEMA. n.d. "Emergency Procedures for Employees with Disabilities in Office Occupancies." U.S. Fire Administration. Retrieved August 17, 2016 from:www.eadassociates.com/fa-154.pdf

Ferleger, David. 2010. "Disabilities and the Law: The Evolution of Independence." *Federal Lawyer*, 57: 26–50.

Finch, Christina, Christopher T. Emrich, and Susan L. Cutter. 2010. "Disaster Disparities and Differential Recovery in New Orleans." *Population & Environment*, 31: 179-202.

Florida Statutes. 2015. "Registry of Persons with Special Needs; Notice; Registration Program, 252.355."

The Florida Legislature. 2015. "252.355 Registry of Persons with Special Needs: Notice; Registration program." Retrieved 5 December, 2015 from: www.leg.state.fl.us/Statutes/index.cfm?App_mode=Display_Statute&Search_String=&URL=0200-0299/0252/Sections/0252.355.html

Fothergill, Alice, and Lori A. Peek. 2004. "Poverty and Disasters in the United States: A Review of Recent Sociological Findings." *Natural Hazards*, 32: 89–110.

Freeberg, Ernest. 2000. "The Meanings of Blindness in Nineteenth-Century America." *Proceedings Of The American Antiquarian Society*, 110: 119–152.

Frieden, Lex. 2006. *The Impact of Hurricanes Katrina and Rita on People with Disabilities: A Look Back and Remaining Challenges*. Washington, D.C.: National Council on Disability.

Fox, Michael H., Glen W. White, Catherine Rooney, and Jennifer L. Rowland. 2007. "Disaster Preparedness and Response for Persons with Mobility Impairments." *Journal of Disability Policy Studies* 17: 196–205.

Hewitt, Paul. 2013. "Organizational Networks and Emergence during Disaster Preparedness: The Case of an Emergency Assistance Registry." Oklahoma State University, 2013. Retrieved 5 December, 2015 from: https://shareok.org/bitstream/han.d.le/11244/10966/Hewett_okstate_0664D_12684.pdf?sequence=1

Hoffman, Sharona. 2009. "Preparing For Disaster: Protecting the Most Vulnerable in Emergencies." *University of California at Davis Law Review*, 42: 1491–1547.

Houston, Nancy, A. Vann Easton, E. A. Davis, J. Minicin, B. D. Phillips, and M. Leckner. 2009. "Evacuating Populations with Special Needs: Routes to Effective Evacuation Planning Primer Series." Washington, D.C.: The Federal Highway Administration, U.S. Department of Transportation.

Jensen, John. 1997. "Before the Surgeon General: Marine Hospitals in Mid-19th-Century America." *Public Health Reports*, 11: 525–527.

Kailes, June Isaacson, and Alexandra Enders. 2007. "Moving Beyond 'Special Needs.'" *Journal of Disability Policy Studies* 17: 230–237.

Katrina Timeline. 2005–2008. Retrieved 5 December, 2015 from: http://thinkprogress.org/katrina-timeline/

Longmore, Paul K. 2009. "Making Disability an Essential Part of American History." *OAH Magazine Of History*, 23: 11–15.

Mann, Stacey. 2014. "Human Resources and Emergency Planning: Preparing Local Governments for Times of Crisis." *Public Administration Quarterly*, 38: 163–205.

Martinko, Mark J., Denise M. Breaux, Arthur D. Martinez, James Summers, and Paul Harvey. 2009. "Hurricane Katrina and Attributions of Responsibility." *Organizational Dynamics*, 38: 52–63.

Miami Dade County. 2015. "About the Emergency & Evacuation Assistance Program." Retrieved 5 December, 2015 from: www.miamidade.gov/hurricane/evacuation-assistance.asp

Morris, Jr., Frank C. 2011. "ADA Amendments: Final EEOC Regulations—What Employers Need to Know." *Insurance Advocate*, 122: 6.

Mullin, Joe. 2015. "9th circuit Rules Netflix Isn't Subject to Disability Law."*Ars Technica*. Retrieved August 23, 2016 from: http://arstechnica.com/tech-policy/2015/04/9th-circuit-rules-netflix-isnt-subject-to-disability-law/

The National Consortium on Leadership and Disability for Youth (NCLD). 2007. "Disability History Timeline: Resource and Discussion Guide." Washington, D.C.: Institute for Educational Leadership.

New York Times. 2008. "Never Again, Again." Editorial, September 20.

Nolan, Bruce. 2005. Katrina takes aim. *Times-Picayune*, August 28. National, p. 1.

Orange County. 2015. "Orange County People with Special Needs Program: Frequently Asked Questions." Retrieved August 17, 2016 from: www.orangecountyfl.net/Portals/0/Library/Emergency-Safety/docs/SpecialNeedsProgramFAQ.pdf

Osceola County. n.d. "Special Needs Registry." Retrieved 5 December, 2015 from: www.osceola.org/search.stml?q=special+needs+registry

Pacer Center. 2004. "History of Disability in Brief." Retrieved May 24, 2016 from: www.pacer.org/C3/curriculum/session1/handouts/History%20of%20Disability%20in%20Brief.pdf

Pasco County Office of Emergency Management. 2013. "Special Needs Information." Retrieved 5 December, 2015 from: www.pascocountyfl.net/DocumentCenter/Home/View/730

Polanco, Elvis. "Office Building Emergency Action Plans: Simplified Analysis of NYC Local Law 26 of 2004." *Radiant Training and Consulting*. Retrieved 5 December, 2015 from: www.radianttraining.com/articles/eap_article-2010_jan.pdf

Pope, John, and Michelle Krupa. 2008. "Storm Plans Put into Motion Already." August 28. *Times-Picayune*, p. 1.

Renne, John L., Pamela Jenkins, Thomas W. Sanchez, and Robert C. Peterson. 2008. The national study on Carless and Special Needs Evacuation Planning: Government and Non-profit Focus Group Results." The University of New Orleans Transportation Center. Retrieved 5 December, 2015 from: www.uno.edu/cola/planning-and-urban-studies/docs/Focus_Group_Final_Report.pdf

Robinson, Eugene. 2005. "Beyond Contrition." *The Washington Post*, 18 September 2005. p. B07.

Santora, Mark, and Benjamin Weiser. Court says New York Neglected Disabled in

Emergencies. 2013. *The New York Times*. Retrieved 5 December, 2015 from: www.nytimes.com/2013/11/08/nyregion/new-yorks-emergency-plans-violate-disabilities-act-judge-says.html?_r=0

Schleifstein, Mark, and Michelle Krupa. "Gustav Has State on Alert; Evacuations Could Start as Early as Friday." *Times-Picayune*. August 27 2008. p. 1.

Schwab, Anna K., and Timothy Beatley. 2013. "Ethics in Catastrophe Readiness and Response." Ed. R. Bissell. *Preparedness and Response for Catastrophic Disasters*. Boca Raton: CRC Press, 45–76.

Society for Human Resource Development. 2014. Disability Accommodations: Conditions: Does the Americans with Disability Act Provide a List of Conditions that are Covered Un.d.er the Act? HR Q & A." Retrieved August 17, 2016 from: www.ada.gov/qandaeng.htm

Special Needs Shelter Interagency Committee, F. R. C. (2012). *State of Florida Functional Needs Support Services Resource Assessment*. Florida Department of Health. Retrieved December 5, 2015 from: www.floridadisaster.org/documents/Florida-FNSS%20Resource%20Assessment%20Report_FINAL.pdf

Stoughton, Linda M. 2015. "Special Needs Working Group 2015." PowerPoint. Retrieved 5 December, 2015 from: www.fepa.org/~fepaorg/attachments/article/453/Mid%20Year%20Powerpoint%20Special%20Needs%20Presentation.pdf

U.S. Court of Appeals. 2011. Baughman v. Walt Disney World Company.

U.S. Department of Education. 2004. "Title I – Improving the Academic Achievement of the Disadvantaged. Retrieved 5 December, 2015 from: www2.ed.gov/policy/elsec/leg/esea02/pg1.html#sec1001

U.S. Department of Health and Human Services. 2014. "Bulletin: HIPAA Privacy in Emergency Situations." Office for Civil Rights.

U.S. Department of Justice. 1998. "U.S. vs. Transco Distributors."

U.S. Department of Justice. 1999. "U.S. vs. Cinemark, USA, Inc." Retrieved 5 December, 2015 from: www.justice.gov/crt/united-states-district-court-northern-district-ohio-eastern-division

U.S. Department of Justice. 2009. "Americans With Disabilities Act of 1990, As Amended, Sec. 12102. Definition of disability."

U.S. Department of Justice. 2009. "Guide to Disability Rights Laws; Americans with Disabilities Act."

U.S. Department of Labor. 2015. "Table 6-A. Employment Status of the Civilian Population by Sex, Age, Disability Status, Not Seasonally Adjusted." Retrieved 5 December, 2015 from: www.bls.gov/news.release/empsit.t06.htm.

United States Senate. 2006. Report of the Senate Committee on Homeland Security and.Governmental Affairs. *Hurricane Katrina: A Nation Still Unprepared*. U.S. Senate Report 109-322, 732 pages. Retrieved August 17, 2016 from www.npr.org/documents/2006/apr/katrina/execsummary.pdf

University Herald. 2014. "PSU to Pay $161,500 to Settle Discrimination Lawsuit Filed by Deaf Student." *University Herald*. Retrieved December 5, 2015 from: www.universityherald.com/articles/7543/20140215/psu-pay-161-500-settle-discrimination-lawsuit-filed-deaf-student.htm

Vogt Sorensen, Barbara. 2006. "Populations with Special Needs." Oak Ridge National Laboratory. Retrieved August 17, 2016 from: www.dhhr.wv.gov/healthprep/about/

archives/Documents/ORISE%20PopulationSpecialNeeds%202006.pdf
Watanabe, L. 2015. "DOJ to Amtrak: Your Train Stations Violate ADA." *Mobility Management*. Retrieved August 23, 2016 from: https://mobilitymgmt.com/articles/2015/06/16/amtrak-ada-violations.aspx
Weber, L. 2015. "Are You Disabled? Your Boss Needs to Know." *Wall Street Journal*. Retrieved December 5, 2015 from: www.wsj.com/articles/SB10001424052702303287804579447450295914372
Welch, Polly. 1995. "A Brief History of Disability Rights Legislation in the United States." *Universal Design Education*. Retrieved May 25, 2016 from: www.udeducation.org/resources/61.html
"A Wet Run." 2008. *Times-Picayune*, August 26. p. 6.
Whoriskey, Peter. 2008. "Jindal Presents a Face of Calm During the Storm; La. Governor Hailed for Recovery Efforts." *The Washington Post*, 3 September. p. A06.
Wolshon, B. 2009. "Transportation's Role in Emergency Evacuation and Reentry. National Cooperative Highway Research Program." Transportation Research Board. Retrieved 5 December, 2015 from: http://onlinepubs.trb.org/onlinepubs/nchrp/nchrp_syn_392.pdf
World Wide Web Consortium. 2008. "Web Content Accessibility Guidelines." Retrieved 5 December, 2015 from: www.w3.org/TR/WCAG20/

Section II
Policy

5 Local Recovery

How Robust Community Rebound Necessarily Comes from the Bottom Up

Emily Chamlee-Wright, Stefanie Haeffele-Balch, and Virgil Henry Storr

Introduction

Hurricanes, tsunamis, earthquakes, tornadoes and other natural disasters cause large-scale physical and emotional damage. In 2005, Hurricane Katrina caused over 1,800 deaths and 125 billion dollars in damage, and displaced more than 400,000 residents (Geaghan 2011). In 2011, an outbreak of tornadoes devastated Tuscaloosa, Alabama and Joplin, Missouri, causing 222 deaths and 5.3 billion dollars in damage.[1] Hurricane Sandy, which affected numerous countries in the Caribbean and every state along the east coast in 2012, caused 285 deaths and 68 billion dollars in damage in the U.S. alone. In less developed countries, the impact of natural disasters is even more pronounced. In 2010, an earthquake of 7.0 magnitude struck Haiti, causing more than 160,000 deaths and as much as 13.2 billion dollars in damage. Many Haitians lost everything. And, with more than a million people living in tent villages after the earthquake, safety and health issues, including a cholera outbreak, have exacerbated the problems of recovery.[2]

Such damage occurs with little to no warning and eliminates many of the systems and structures of support that make the routines of daily life possible. Immediately after a storm, residents may be without power, water and food. If they evacuated, they may be separated from loved ones and unable to determine if and when they can return. With housing and workplaces damaged or destroyed, there is uncertainty about whether to return and rebuild or to move and make a fresh start in a different location. Such difficult decisions often depend on the decisions and actions of one's family, friends and neighbors. It is only worthwhile to return and rebuild if others return as well. In such times of uncertainty, it is in everyone's incentive to wait for others to act first. This "collective-action problem" is inherent in the post-disaster context (Chamlee-Wright and Storr 2009a, 2009b, 2010a).

The financial resources, time and effort required to repair and rebuild damaged housing, businesses and public infrastructure far surpass the resources available at the local level. The magnitude of the destruction and the resources needed for recovery are so vast that the necessity for federal government intervention to fund and coordinate disaster response and recovery is more often than not simply assumed (Birch and Wachter 2006; Cigler 2009, pp. 759–766; Springer 2009, p. 10). Governments, with their ready access to personnel and resources, have grown to play a major role in emergency response, cleanup and community redevelopment planning.

However, the collective-action problem associated with the post-disaster context also affects government action. Just as citizens require government action post-disaster, particularly in the delivery of key public goods, government requires citizens to act in order to determine how best to prioritize and direct those services. In other words, it is impossible to assess which neighborhoods should get first priority if all residents are waiting on the sidelines. Determining where to begin clearing debris, restoring power and reopening public schools, for example, is not obvious; residents often must return and then demand that services be restored.[3]

In the post-disaster context, it is often local entrepreneurs and community leaders who signal others to return and encourage recovery. It is the local minister or rabbi who contacts his parishioners to make sure they are safe, who turns the church parking lot into a temporary relief center, who turns the Sunday school classrooms into a makeshift daycare and who preaches to an empty hall until residents begin returning. It is the local businesswoman who takes a risk by returning and reopening her store, hoping that others will soon come back. In some cases businesses provide goods essential to the rebuilding process, such as construction materials. In other cases, reopened businesses provide essential services, such as ready-to-eat meals, to returnees who have no other source of food or facilities in which to cook. In still other cases, the services businesses provide may at first blush seem nonessential, but their return is a signal that normal life in the community is on the horizon (Chamlee-Wright 2008, pp. 615–626). Through these actions, they provide the answers that others are waiting for in order to start rebuilding their community.

After Hurricane Katrina struck the Gulf Coast in 2005, a team of scholars, including two of the authors of this chapter, embarked on a five-year project to examine community rebound after the storm. During that time, the Gulf Coast Recovery Project team conducted more than 300 interviews with residents in New Orleans, Louisiana and more than 100 surveys and/or interviews with displaced residents in Houston, Texas (Chamlee-Wright 2010). The result of this effort was a body of research on community recovery, including examination of the role of local entrepreneurs, church leaders, businesses, nonprofits and governments. Since then, additional research, following the interview structure and research strategy of the Gulf Coast Recovery Project, has been conducted following the 2011 tornado outbreak in Tuscaloosa, Alabama and Joplin, Missouri, as well as in Rockaway, New York after Hurricane Sandy in 2012.

The common theme across these interviews, locations and disasters has been the importance of local leaders and decentralized organizations in spurring community rebound. This characteristic was found both in close-knit homogenous communities, such as the neighborhood surrounding the Mary Queen of Vietnam Catholic Church in East New Orleans, and diverse communities, such as the Broadmoor neighborhood in central New Orleans (Chamlee-Wright and Storr 2009b; Chamlee-Wright and Storr 2010a; Storr and Haeffele-Balch 2012, pp. 295–314). When local leaders and entrepreneurs return and signal that others can (and should) return as well, they spur recovery and help to set their communities back on track. They succeed by (a) accessing and reconnecting social networks within the community, (b) providing goods and services that residents need in order to return, (c) advocating for their community in the redevelopment planning and recovery process and (d) navigating the insurance and assistance process in order to access resources for their community. It is our contention that this common occurrence is *a necessary force* for post-disaster community rebound.

The federal government does formally recognize the importance of community involvement in disaster-preparedness, response and recovery. Speaking of the importance of "engaging the Whole Community in preparedness activities," for instance, Federal Emergency Management Agency (FEMA) Administrator Craig Fugate (Fugate 2011, n.p.) notes the need to "engage with partners at every level of government as well as the nonprofit and private sector." Attempts to incorporate this Whole Community approach include inviting representatives from nonprofit organizations and businesses to join planning efforts and to sit in at the operation centers during disasters. However, such efforts to incorporate local representatives into federal intervention fail to capture the benefits of community-led rebound: it is the presence of local entrepreneurs and community members on the ground, reacting and adapting to specific needs and circumstances, that brings about recovery.

This chapter argues that swifter and more robust community rebound can be fostered not by government "engaging" local actors in the top-down system of disaster management, and not by government mimicking the best practices of local entrepreneurs and stakeholders, but by transferring actual decision authority away from federal, state and even municipal government actors and agencies and toward local communities. This chapter proceeds as follows: the second section examines the challenges inherent in federal, top-down disaster assistance and analyzes recent efforts to incorporate local involvement into federal disaster assistance, specifically through the Post-Katrina Emergency Management Reform Act of 2006; the third section examines the literature on the necessity of decentralized and local efforts in community rebound, providing examples from Hurricane Katrina and the tornadoes in Joplin, Missouri; and the fourth section provides policy implications and concludes.

Disaster Assistance from the Top Down

Following a disaster, the government faces the same collective-action problem confronted by individuals when trying to coordinate the activities of hundreds of thousands of displaced residents and to build community confidence that recovery will succeed. Just as individuals have an incentive to wait until others return, the government may also wait for residents to return in order to determine which neighborhoods should get priority when restoring utilities, clearing roads and repairing infrastructure. In the case of widespread disaster, the collective-action problem that governments face is amplified by the fact that information is dispersed among individuals across multiple smaller communities and must be somehow gleaned and integrated in order to develop a successful post-disaster redevelopment plan. The fact that disasters often trigger large-scale displacement to various and disconnected locales exacerbates the challenge of locating and communicating with residents to determine their plans to return and rebuild (or not) and the resources and conditions needed for them to succeed. In other words, government faces what some within the economics profession have called "the knowledge problem." Hayek (1945, pp. 519–530) identified and articulated this knowledge problem in his essay "The Use of Knowledge in Society." Hayek argued that markets, through the use of prices and the profit-and-loss system, direct resources to their highest valued use despite information being dispersed and context-dependent. Central planners, however, cannot gain access to and use the knowledge of society in the same way that individuals and organizations can within markets. Speaking of top-down government control over the economy, Hayek (Ibid, p. 520) observed that central planners face "a problem of the utilization of knowledge not given to anyone in its totality."

In "The Use of Knowledge in Natural-Disaster Relief Management," Sobel and Leeson (2007) apply the knowledge problem to the post-disaster situation and argue that, if disaster management is to be effective, government agencies must be able to (a) recognize that a crisis has occurred, (b) determine what needs must be met and how supplies and resources will be allocated to best serve those needs and (c) evaluate and modify the relief effort as circumstances change. Looking at immediate post-Katrina relief efforts, they concluded that the government agencies tasked with disaster-relief management failed on multiple levels. First, the Department of Homeland Security was late in declaring Hurricane Katrina a disaster; second, FEMA misallocated and often failed to allocate labor and resources that could and should have been used in the relief effort; and third, even though government actions were widely criticized, FEMA received a larger budget and more responsibility for the federal government's post-Katrina response (Ibid).

With regard to disaster recovery, Storr and Haeffele-Balch (2012, p. 320) argue that

> If a government is to be effective at planning and leading recovery, it must determine (a) which residents are most likely to return and which neighborhoods are most likely to rebound, (b) how best to allocate resources, and (c) when it has made mistakes and how to correct them.

Again, Hurricane Katrina highlighted the difficulties faced by government in the post-disaster situation. For instance, the city of New Orleans discussed five different recovery plans in three years after Hurricane Katrina, each one focusing on rebuilding and investing in different areas (Olshansky, Johnson, Horne and Nee 2008). Such fluctuations in the planning process add additional uncertainty to an already difficult situation, resulting in wasted time and resources as citizens wait for final plans or adjust to changing rules of the rebuilding "game." Furthermore, community leaders and residents faced challenges when trying to access resources, get permits and navigate the assistance process. Local leaders, such as the pastors and staff of the Mary Queen of Vietnam Catholic Church in New Orleans East, necessarily had to become adept at filling out applications, writing grant proposals and lobbying local and federal politicians to get resources and the permission needed to rebuild their community (Chamlee-Wright 2011, pp. 167–185). These examples address some of the difficulties governments face when attempting to determine (a) which residents and neighborhoods will return and recover and (b) how to best allocate resources.

In response to the myriad of response and recovery issues following Hurricane Katrina, Congress passed the Post-Katrina Emergency Management Reform Act of 2006 (PKEMRA) in order to give FEMA the authority and guidance to improve their efforts. Among the changes, PKEMRA directs FEMA to provide technical assistance to local and state governments to develop plans for disaster response and recovery, increase the nation's communication interoperability and ensure the nation's disaster-preparedness by assessing, assisting and standardizing local, regional and federal planning efforts. In many ways, FEMA was tasked not only with improving its own efforts but with ensuring that citizens and all levels of government are prepared for disasters.

Five years after the passage of PKEMRA, FEMA Administrator Craig Fugate testified on the importance of the Act and the need to involve the "Whole Community" in the preparedness and recovery processes:

> For the first time, it gave FEMA clear guidance on its mission and priorities, and provided us with the authorities and tools we needed to become a more effective and efficient agency, and a better partner to state, local, territorial, and tribal governments.
> (Fugate 2011, p. 2)

And,

> In particular, we have made significant improvements to our approach to preparedness. We now focus on engaging the Whole Community in preparedness activities. We have realized that a federal-centric approach will not yield success and that instead we must collaborate and engage with partners at every level of government as well as the nonprofit and private sector. But there is more work to be accomplished.
> (Fugate 2011, p. 2)

Fugate went on to summarize the partnerships FEMA has forged with state and local emergency managers, including strategy meetings with the Red Cross and other voluntary organizations and the drafting of documentations to serve as guidelines and credentialing tools for national preparedness, response and recovery. Throughout the testimony, Fugate emphasized the need for FEMA to coordinate and oversee disaster-preparedness and recovery efforts, to centralize the "Whole Community."

While FEMA and other government entities can play an important role in coordinating intergovernmental discussions and advising communities on preparedness, much of the public policy reforms have focused on preparedness rather than recovery and on advancing national goals rather than those of the community. For instance, the 2011 Presidential Policy Directive on National Preparedness (PPD-8) tasked the Department of Homeland Security and FEMA to develop a national preparedness system, including developing operational plans and prioritizing capabilities and skills needed for responding to disasters and other risks. The policy literature emphasizes the need for centralization as well, calling for full integration and a unified command structure with a strong FEMA taking the lead (Cigler 2009, pp. 19–26; Peters 2008, pp. 19–26; Springer 2009; Tierney 2007).

The passage and implementation of PKEMRA does show that FEMA and the federal government recognized that mistakes had been made (part of the third determinate of effective government-led recovery). FEMA administrators have widely recognized that all disasters are essentially local in nature (Fugate 2011; Pittman 2011). However, calls to "keep it local" have largely been overlooked by government entities. Instead, public policy has focused on nation-wide preparedness goals and metrics, and unified command after a disaster. As Crabill and Rademacher (2012) observed,

> Our historical response has been to draft policy that reacts to the issues that surfaced during the disaster response, expanding the federal role and thus driving the message that "We will save you next time." The legislation, programs and policies instituted post-disaster have virtually ignored the potential of further engaging the individual, local and state role. Instead they have perpetuated further dependency on the federal government.

Crabill and Rademacher attribute the continued challenges to three underlying causes: (a) the government's roles and fiscal responsibilities outlined in legislation do not match actual capabilities; (b) federal programs actually create dependency and increase risk; and (c) the federal government continues to assume and declare that it has the capabilities to coordinate and direct disaster response and recovery. Such efforts continue to emphasize the need for central planners to oversee disaster management, rather than promoting the role of communities in preparing for and managing rebound after disasters.

Recovery from the bottom up

Following a disaster, communities have limited access to resources and face uncertainty over when those resources will arrive and services will be restored. The pre-existing connections and relationships among community members can play a crucial part in overcoming these challenges and encouraging recovery (Chamlee-Wright 2010). These connections, or social capital, are how groups of individuals interact in efforts to achieve their individual and collective goals (Bourdieu 1985; Coleman 1988; Adler and Kwon 2002). Social capital has been found to be correlated with educational attainment, financial success and economic growth (meaning that groups can improve the wellbeing of their individual members by cultivating social capital), both in mundane times as well as in times of crisis (Coleman 1988; Granovetter 1973; Knack and Keefer 1997; Woolcock 1998).

Woolcock (2001) identified three types of social capital: bonding, bridging and linking. Bonding social capital is found in close-knit and homogenous groups, such as the bonds of a church group. Bridging social capital is found in more loose-knit and heterogeneous groups, such as the relationships formed at work or at school. And finally, linking social capital exists when relationships are made across groups, such as the connections made through networking. These different types of relationships are the ways in which individuals interact with others in their community, and provide channels for social learning and widespread communication. For instance, individuals can imitate the actions of others in the network, signaling which actions and decisions are successful and which are not (Burt 2001, pp. 31–56).

In the post-disaster context, communities can share information, resources and successful methods for response and recovery through these various types of social capital. For instance, Chamlee-Wright and Storr (2009b) show how the immigrant community associated with the Mary Queen of Vietnam Catholic Church in New Orleans East made a rapid return and recovery after Hurricane Katrina through the leadership of their head pastor, Father Vien. Similarly, Hurlbert *et al.* (Hurlbert, Haines and Beggs 2000; Hurlbert, Haines and Beggs 2001) and Aldrich (Aldrich 2012; Aldrich 2011a; Aldrich 2011b) highlight the importance of bonding social capital in post-hurricane and tsunami recovery. Bridging social capital, present in more diverse communities with weaker social ties like that of the Broadmoor neighborhood in central New Orleans, was also successfully utilized to coordinate plans and encourage community rebound (Chamlee-Wright and Storr 2008 and 2009b; Storr and Haeffele-Balch 2012). Furthermore, Hawkins and Maurer (2010, pp. 1777–1793) examined how low-income families utilized all three types of social capital to rebuild their lives after Hurricane Katrina, highlighting that both bridging and linking social capital were important for long-term survival and community revitalization.

By accessing pre-existing social capital and by adapting the operations of pre-existing organizations to the requirements of a post-disaster context, entrepreneurs, local leaders and for-profit and nonprofit organizations can encourage return and recovery after a disaster. They utilize their prior experience, relationships and skills to respond to the new and ever-changing needs of their community in order to (a) contact displaced residents, encouraging them to return and connecting them to the networks of emotional and material support they need to return, (b) provide goods and services that residents need in order to return, (c) advocate for their community and influence the redevelopment planning and recovery process and (d) navigate the insurance and assistance process in order to access resources for their community. Such community leaders show up again and again in our research after Hurricane Katrina, the 2011 tornado outbreak and Hurricane Sandy. Below we offer a few examples, highlighting the common theme of the importance of community leaders in disaster recovery.

Father Vien, the pastor of the Mary Queen of Vietnam Catholic Church when Hurricane Katrina struck New Orleans, was adept at all four of the aspects of community rebound mentioned above. After the storm, Father Vien made personal visits to parishioners scattered across evacuation sites in multiple states to encourage them to come home, and began holding church services within six weeks after the storm (Chamlee-Wright and Storr 2010a). Father Vien also spoke to news outlets about the vitality and resilience of his community, and when faced with bureaucratic red tape, he used his access to the media to ensure that his community avoided bureaucratic roadblocks and attract resources to the community. For instance, he led an effort to obtain FEMA trailers for the community and set them up on church grounds. When the city refused to fulfill the permit, Father Vien went to City Council meetings, contacted the media, petitioned politicians for support and eventually was granted permission to open the trailer park (Chamlee-Wright and Storr 2011). He also convinced Entergy, the local power company, that services should be restored in his neighborhood by showing photographic evidence as well as the names of all the residents who had returned and attended church services (Chamlee-Wright and Storr 2010a).

Similarly, in the Broadmoor neighborhood of New Orleans, LaToya Cantrell became a rallying force and spokesperson for her community in the aftermath of the storm. In early 2006, displaced residents were shocked to learn that the Bring New Orleans Back Commission was planning to turn their neighborhood into green space unless they could prove that 50 percent were committed to return. Cantrell, as president of the Broadmoor Improvement Association, led the effort to contact residents to encourage them to return. Broadmoor residents used their expertise and contacts to build a media and promotion campaign, enlist help from churches and nonprofit organizations, apply for grants and lobby for community development projects (Storr and Haeffele-Balch 2012). Their efforts not only ensured that every additional redevelopment plan included Broadmoor, but also brought additional attention and resources into the community.

Across town, Doris Voitier took up the task of spurring recovery in St. Bernard Parish. As the superintendent of the school district, Voitier knew that first responders needed a functioning school for their children to attend and pledged that there would be space for every child who registered on November 1, 2005. Over 700 students registered that day. Chamlee-Wright and Storr (2010a, p. 156), noted:

> Many of our interviewees noted that they could not contemplate returning to New Orleans until the schools reopened. It can be difficult, however, for officials to commit scarce resources to schools that do not yet have students. By committing to reopen the schools on a particular date, Voitier signaled to displaced residents that the community recovery would, indeed, occur and allowed residents to make clear decisions about whether and when to return.

In order to follow through on her promise, Voitier had to find ways to circumvent the traditional process of rebuilding a public school. For example, when she asked for trailers to serve as temporary classrooms from the Army Corps of Engineers, she was told they would not be ready until the next academic year. Instead, she worked with a local contractor and purchased them herself. As Chamlee-Wright and Storr concluded, "Although Voitier is a government official, she had to work (as a social entrepreneur) outside the government sector in order to honor her commitment" (2010a, p. 156).

Additionally, business owners played a major role in spurring return and providing goods and services needed for recovery. After Hurricane Katrina, Ben Cicek opened a coffee shop in St. Bernard Parish to provide coffee, meals and internet access for residents and construction workers. By doing so, he also established a social space where people could discuss their troubles and apply for assistance. In Joplin, Missouri, Tony Calderone, a local hardware store manager, responded quickly to the needs of the community by handing out water, hammers and other small items to residents, as well as supplying needed equipment and resources. David Glenn, a commercial real estate broker in Joplin, went back to work immediately following the tornado outbreak in order to find temporary commercial real estate for the local businesses that sustained damage, as well as the influx of contractors, insurance companies and other businesses aiding in recovery (Smith and Sutter 2013, pp. 165–188).

Community leaders and local entrepreneurs—such as Father Vien, Latoya Cantrell, Doris Voitier, Ben Cicek, Tony Calderone and David Glenn—play a key role in spurring community rebound. Their success is based on their personal experiences and connections as well as local, contextual knowledge of their community's needs; it is their presence and actions on the ground post-disaster that lead to successful rebound. Any attempt to generalize these successes in order to incorporate them into government response and recovery efforts will be devoid of the essential characteristics of local action (Chamlee-Wright and Storr 2011).[4]

Policy Implications and Conclusion

Despite the recognition that community involvement is important to recovery, FEMA's effort to incorporate communities into disaster planning, response and recovery have missed the point of community-led rebound. Mimicking best practices that were discovered on the ground after a disaster will not enable communities to adapt to the inevitable differences each disaster brings. Policies that encourage community-led recovery will ease barriers to rebuilding and recovery. In order to fully incorporate the lessons from successful community rebound, government entities should (a) transfer decision authority to local communities, (b) anchor community expectations around achievable positive outcomes and (c) focus on providing the basics in the immediate aftermath of a disaster.

Communities are built by residents, business owners and leaders of local charities, not by governments. And as the research on community rebound shows, communities are rebuilt by those individuals as well. Government agencies and government-appointed committees tasked with redevelopment planning are often tempted by the notion that the post-disaster moment is the time to reengineer a city, to build it back better and smarter than before. But such ambitions are worse than unrealistic; they do real damage, in that they delay rebuilding in anticipation of the perfect plan. Such delays reinforce the collective action problem by anchoring negative expectations in the minds of those waiting on the sidelines for signs of recovery before they return (Chamlee-Wright and Storr 2009a). As Gordon and Ikeda (2009) state, there are always limits to urban planning, and those limits are only exacerbated during a crisis.

Instead of initiating a major planning effort after a disaster, government entities should transfer decision authority to local communities and neighborhoods, and clearly signal that the authority that local communities and private citizens possessed pre-disaster is maintained post-disaster. Instead of waiting for approval to rebuild from official plans or spending time petitioning to change and influence the planning process, residents should be free to rebuild or develop new businesses based on their own plans and goals. Every day spent planning is a day spent not rebuilding. For instance, many of the most successful rebuilding efforts after Hurricane Katrina, such as the Vietnamese community in New Orleans East and the Broadmoor neighborhood in central New Orleans, would have been prohibited by the Bring New Orleans Back Commission at the beginning of 2006. Only after these communities petitioned and proved their vitality were they incorporated into the recovery plans. This exemplifies the fact that planners cannot predict which communities will succeed or fail.[5]

Similarly, government entities should focus on anchoring expectations around realistic and achievable outcomes. As Chamlee-Wright and Storr (2010b) found, though residents are often pessimistic about the *intentions* of government actors, they tend to be optimistic about the *capabilities* of government to provide effective post-disaster assistance. In other words, people tend to be optimistic about government's ability to solve all manner of post-disaster problems, and it is only a matter of political will, most people believe, that stands in the way. The actual capabilities of government, however, are limited. By promising only what is absolutely necessary and achievable, policymakers foster certainty about what kinds of assistance are indeed on their way, thereby fostering more effective planning at the individual and local community level. Such realistic expectations and local planning, in turn, foster community-led rebound. Further, by transferring decision authority back to local communities, policymakers reduce the confusion associated with regulatory and bureaucratic rigidity and uncertainty regarding the rules governing the rebuilding process associated with top-down redevelopment planning (Chamlee-Wright 2007, 2010). Again, scaling back post-disaster government action and anchoring expectations accordingly will allow for community-led recovery. As Crabill and Rademacher note, "streamlining the federal role and communicating what it can and cannot do will increase public confidence over the long term" (Crabill and Rademacher 2012).

Therefore, when it comes to official disaster response and assistance, government entities should promise relatively little—such as restoring electricity, water and other utilities, as well as clearing debris—and deliver on those promises swiftly. Public policy can help anchor community expectations around the likelihood of a successful rebound, not by promising to reengineer the community from the top down, but by clearing debris and just getting needed services restored. After these basic services are restored, local, state and federal government can turn its attention to repairing infrastructure, such as hospitals and schools. Additionally, disaster assistance will be more effective if it comes in a form that allows individuals to use that assistance in a manner that is most appropriate for their circumstances. Rather than providing temporary trailers, for example, quickly putting housing vouchers in the hands of residents would allow them to find the temporary housing that is most appropriate to their needs (Chamlee-Wright 2007).

Last, government entities should prepare for which rules and regulations should be enforced and which ones should be relaxed after a disaster. Upholding and enforcing the rule of law, private property rights and contracts will allow residents, businesses and charities to go about recovery efforts without fear that the basic rules that govern the social order are in doubt. However, rigid rules regarding occupational licencing—for example that of contractors, daycare providers and so on—that may be deemed appropriate in ordinary circumstances can significantly limit the inflow of critical service providers from the for-profit and nonprofit sector in a post-disaster context. Similarly, regulator controls designed to attend to issues of environmental or historical preservation, which seem reasonable in general, can present critical roadblocks post-disaster. Having a clear and credible plan to relax such regulations and laws, without disrespecting private property rights, before a disaster occurs can help spur rebuilding in a post-disaster situation.

By focusing on reforms that scale back the role of government and encouraging public policy that enables community-led rebuilding, government entities can reduce some of the uncertainty and burden of disaster recovery. Such efforts will lessen the collective action problem inherent in the post-disaster context and provide space for community leaders and entrepreneurs to spur community rebound.

Notes

1 The tornado in Tuscaloosa, Alabama caused 64 deaths and $2.5 billion in property damage and the tornado in Joplin, Missouri caused 158 deaths and $2.8 billion in property damage. For more information, see www.spc.noaa.gov/climo/torn/2011deadlytorn.html and www.spc.noaa.gov/faq/tornado/damage$.htm.
2 For more information, see www.cnn.com/2012/10/18/world/americas/cnnheroes-haiti-rape/ and www.cdc.gov/haiticholera/haiti_cholera.htm.
3 While after some disasters, the locations that are hit the hardest may be relatively small and easy to determine, response agencies and municipal utility companies still face issues determining how to prioritize restoring services.
4 Furthermore, while FEMA and other government agencies may be able to identify local leaders and incorporate them into the larger response and recovery efforts, actively including individuals and private organizations will incentivize others to seek the benefits of federal support. As more people and organizations seek involvement, government agencies will find it increasingly difficult to determine who has local knowledge that will be useful to the recovery process and who is simply rent-seeking. Recognizing that the lines between a highly engaged community leader/organization possessing significant local knowledge, and a person or organization that is engaging in an act of pure rent seeking will always be blurred, any interaction involving transfer of resources should be extremely local and specific. For more information on how local leaders accumulate lobbying social capital when interacting with government agencies, see Chamlee-Wright and Storr (2011).
5 An argument could be made for the need for city planners to assess whether residents can return and rebuild in risky locations, such as areas prone to flooding. Such problems would be better addressed, however, by removing flood insurance subsidies and rendering insurance premiums actuarially sound pre-disaster, not by picking winners and losers ad hoc post-disaster. By doing so, individual property owners will be in a better position to decide whether to rebuild or to relocate, based on true costs (insurance, repairs, etc.) and benefits (location preferences, a sense of community, a family legacy, etc.).

Bibliography

Adler, P. S. and S.W. Kwon. "Social Capital: Prospects for a New Concept." *The Academy of Management Review* 27, no. 1 (2002): 17–40.

Aldrich, D. "The Power of People: Social Capital's Role in Recovery from the 1995 Kobe Earthquake." *Natural Hazards* 56, no. 3 (2011a): 595–611.

Aldrich, D. "The Externalities of Social Capital: Post-Tsunami Recovery in Southeast India." *Journal of Civil Society* 8, no. 1 (2011b): 81–99.

Aldrich, D. *Building Resilience: Social Capital in Post-Disaster Recovery*. Chicago: University of Chicago Press, 2012.

Birch, E. L. and S. M. Wachter. *Rebuilding Urban Places after Disaster: Lessons from Hurricane Katrina*. Philadelphia: University of Pennsylvania Press, 2006.

Bourdieu, P. "The Forms of Capital." In *Handbook of Theory and Research for the Sociology of Education, edited by* J. G. Richardson, 241–258. New York: Greenwood, 1985.

Burt, R. "Structural Holes versus Network Closure as Social Capital." In *Social Capital: Theory and Research*, edited by N. Lin, K. Cook and R. Burt, 31–56. New York: Aldine De Grutyer, 2001.

Chamlee-Wright, E. "The Long Road Back: Signal Noise in the Post-Katrina Context." *The Independent Review* 12, no. 2 (2007): 235–259.

Chamlee-Wright, E. "Signaling Effects of Commercial and Civil Society in Post-Katrina Reconstruction." *International Journal of Social Economics* 35, no. 8 (2008): 615–626.

Chamlee-Wright, E. *The Cultural and Political Economy of Recovery: Social Learning in a Post-Disaster Environment*. London: Routledge, 2010.

Chamlee-Wright, E. and V. H. Storr. "The Entrepreneur's Role in Post-Disaster Community Recovery: Implications for Post-Disaster Recovery Policy." *Mercatus Center Policy Primer*, Policy Primer No. 6. Arlington, VA: Mercatus Center at George Mason University, September 2008.

Chamlee-Wright, E. and V. H. Storr. "Filling the Civil Society Vacuum: Post-Disaster Policy and Community Response." *Mercatus Center Policy Series*, Policy Comment No. 22. Arlington, VA: Mercatus Center at George Mason University, February, 2009a.

Chamlee-Wright, E. and V. H. Storr. "Club Goods and Post-Disaster Community Return." *Rationality and Society* 21, no. 4 (2009b): 429–458.

Chamlee-Wright, E. and V. H. Storr. "The Role of Social Entrepreneurship in Post-Katrina Community Recovery." *International Journal of Innovation and Regional Development* 2, no. 1/2 (2010a): 149–164.

Chamlee-Wright, E. and V. H. Storr. "Expectations of Government's Response to Disaster." *Public Choice* 144, no. 1 (2010b): 253–274.

Chamlee-Wright, E. and V. H. Storr. "Social Capital, Lobbying and Community-Based Interest Groups." *Public Choice* 149, nos. 1/2 (2011): 167–185.

Cigler, B. A. "Emergency Management Challenges for the Obama Presidency." *International Journal of Public Administration* 32, no. 9 (2009): 759–766.

Coleman, J. S. "Social Capital in the Creation of Human Capital." *American Journal of Sociology* 94 Supplement (1988): S95–S120.

Crabill, A. and Y. Rademacher. "Breaking the Cycle of Reliance on Federal Help After Disasters." *Emergency Management*, May 16, 2012. Accessed February 25, 2016. www.emergencymgmt.com/disaster/Breaking-Reliance-Federal-Help-After-Disasters.html?page=1.

Fugate, C. "Five Years Later: An Assessment of the Post Katrina Emergency Management Reform Act." FEMA Written Statement before the House Committee on Homeland Security, Subcommittee on Emergency Preparedness, Response, and Communications, United States Congress, 112th Cong., 1, Washington, DC, October 25, 2011. www.dhs.gov/news/2011/10/25/written-testimony-fema-house-homeland-security-subcommittee-emergency-preparedness. Accessed 15 August, 2016.

Geaghan, K. A. "Forced to Move: An Analysis of Hurricane Katrine Movers." *SEHSD Working Paper* (2011): 2011–2017.

Gordon, P. and S. Ikeda. "Power to the Neighborhoods: The Devolution of Authority in Post-Katrina New Orleans." *Mercatus Center Policy Series*, Policy Comment No. 12. Arlington, VA: Mercatus Center at George Mason University, February 2009.

Granovetter, M. "The Strength of Weak Ties." *American Journal of Sociology* 78, no. 6 (1973): 1360–1380.

Hawkins, R.L. and K. Maurer. "Bonding, Bridging and Linking: How Social Capital Operated in New Orleans following Hurricane Katrina." *British Journal of Social Work* 40 (2010): 1777–1793.

Hayek, F.A. "The Use of Knowledge in Society." *American Economic Review* 35, no. 4 (1945): 519–530.

Hurlbert, J., V. Haines and J. Beggs. "Core Networks and Tie Activation: What Kinds of Routine Networks Allocate Resources in Nonroutine Situations?" *American Sociological Review* 65, no. 4 (2000): 598–618.

Hurlbert, J., J. Beggs and V. Haines. "Social Capital in Extreme Environments." In *Social Capital: Theory and Research*, edited by N. Lin, K. Cook and R. Burt, 209–231. New York: Aldine De Gruyter, 2001.

Knack, S. and P. Keefer. "Does Social Capital Have an Economic Payoff? A Cross-Country Investigation." *The Quarterly Journal of Economics* 112, no. 4 (1997): 1251–1288.

Olshansky, R.B., L.A. Johnson, J. Horne and B. Nee. "Longer View: Planning for the Rebuilding of New Orleans." *Journal of the American Planning Association* 74, no. 3 (2008): 273–287.

Peters, K. M. "Recovering FEMA." *Government Executive* 40, no. 4 (2008): 19–26.

Pittman, E. "Remember: All Disasters Are Local, Says FEMA Deputy Administrator." *Emergency Management*, November 14, 2011. Accessed February 25, 2016. www.emergencymgmt.com/disaster/Remember-All-Disasters-Are-Local-Says-FEMA-Deputy-Administrator.html.

Smith, D. and D. Sutter. "Response and Recovery after the Joplin Tornado: Lessons Applied and Lessons Learned." *The Independent Review* 18, no. 2 (2013): 165–188.

Sobel, R. and P. Leeson. "The Use of Knowledge in Natural-Disaster Relief Management." *The Independent Review* 11, no. 4 (2007): 519–532.

Springer, C. G. "Strategic Management of Disaster Preparedness." *PA Times* 32, no. 6 (2009): 10.

Springer, C. G. "Achieving Community Preparedness Post-Katrina." *The Public Manager* 39, no. 3 (2010): 42–45.

Storr, V. H. and S. Haeffele-Balch. "Post-Disaster Community Recovery in Heterogeneous, Loosely Connected Communities." *Review of Social Economy* 70, no. 3 (2012): 295–314.

Tierney, K. J. "Testimony on Needed Emergency Management Reforms." *Journal of Homeland Security and Emergency Management* 4, no. 3 (2007). Accessed February 25, 2016. doi:10.2202/1547-7355.1388.

Woolcock, M. "Social Capital and Economic Development: Toward a Theoretical Synthesis and Policy Framework." *Theory and Society* 27 (1998): 151–208.

Woolcock, M. "The Place of Social Capital in Understanding Social and Economic Outcomes." *Canadian Journal of Policy Research* 2, no. 1 (2001): 11–17.

6 Small Businesses as a Vulnerable Population

Mark R. Landahl and Tonya T. Neaves

Introduction

Political figures and economists alike routinely point to small business as a key component of the U.S. national economy. In one recent example, data released by the Bureau of Labor Statistics shows that in 2014, businesses with less than fifty employees accounted for 34 percent of new job growth (BLS 2015). Small businesses contribute in many areas beyond job growth, often described as the "backbone" of the American economy and a cornerstone of the American dream. The importance of small businesses in the local, state, regional and national economies is without question. However, all businesses are not created equal: not in size, sales or scope. This is true both in the normal business operating environment and when disaster strikes.

The disaster literature consists of limited studies with businesses as a unit of analysis. Businesses only recently became the subject of study, compared to traditional elements of society (such as government and household). Tierney (2006) identifies that earlier research generally focuses on impacts at the aggregate community, regional and national levels (Friesma *et al.* 1979; Rossi *et al.* 1978; Thompson 2009). Research interest in businesses and the future impacts of climate change, however, has increased attention on the topic.

This chapter provides a synthesis of existing disaster research on businesses as a unit of analysis across the four recognizable phases of emergency management: preparedness, mitigation, response and recovery. It 1) examines definitional issues for operationalization of small businesses in both government policy and as a unit of analysis in disaster research; 2) considers key findings of academic research concerning small businesses in disaster; and 3) argues for the conceptualization of small businesses as a vulnerable population in disaster. Finally, the chapter discusses implications for local emergency management practice for engaging small businesses in disaster preparedness.

Review of Literature

The emergency management field suffers from a multitude of definitional difficulties. These begin with the fundamental question of the field: *What is a disaster?* The study of businesses presents several additional definitional challenges for disaster research. Questions arise regarding how to classify and study businesses in disaster. Are the legal ownership categories, business sector of operation, number of employees or annual revenues individually or collectively critical for the study of business? For example, many studies either specifically target or show disparities between large and small business. In the limited existing research, a specific difficulty arises in cross-study comparison for the terms "large" and "small" business.

The intent of this chapter is to introduce the complexity of policy and research-based definitions of small business and review existing research on businesses across the phases of emergency management, essentially answering two questions: 1) What is a small business from the perspective of government policy and emergency-management research? and 2) What does research in emergency management tell us about business in terms of the phases of disaster?

The U.S. Policy Perspective: *When is a business large or small or not a business at all?*

From a policy perspective, the moniker *small business* has important implications for access to programs and preferences for taxation and contracting at all levels of government. It is obviously an area of intense political interest, as definitions have financial ramifications for the inclusion or exclusion of a small business from government programs. Policies related to small business reside at multiple levels of government. Federal, state and local governments maintain programs targeted at small businesses. Each program contains definitional criteria for program eligibility. These criteria differ by several metrics; even the identification of business sectors themselves may differ across levels of government, as shown in the example below from the state of Maryland. The policy explanation that follows shows the complexity that goes into the two seemingly simple words: *small* and *business*. The subsequent paragraphs discuss the background of federal-level definitions of small business and highlight a state and a local example, both from the state of Maryland.

The definition of small business is not a static concept. At the federal level, the Small Business Administration (SBA) within the U.S. Department of Commerce develops the definition of small business for access to programs and preferences. The Small Business Act of 1953 empowers the SBA to determine standards for defining small businesses (13-CFR-121).

The regulation does not provide specific standards, but directs the SBA to create and maintain the standards. The SBA creates standards based upon market analysis differentiated by sectors outlined in the North American Industry Classification System (NAICS). As a result, the SBA defines small businesses in 1,141 categories and eighteen sub-industry activities within the NAICS (Dilger 2012). The SBA market analysis results in small-business standards using five metrics: 1) number of employees; 2) average receipts (three previous years); 3) asset size; 4) barrel-per-day refining capacity for petroleum; and 5) megawatt hours for electric power industries (Dilger 2012). The SBA qualifies the small-business designation on analysis of each category in order to meet its legislatively intended mission: to prevent monopolies and promote competition in markets (Dilger 2012).

The characteristics of the specific industry drive the metrics which the SBA uses to define a business as "small." For example, the number of employees that defines a small business is differentiated across industrial classifications. In the air transportation sector, a small business has fewer than 1,500 employees, whereas a retail fuel-sector business is small if it has fewer than fifty employees (SBA n.d.). In addition, the SBA Office of Advocacy broadly defines a "small business" as an independent business with fewer than 500 employees (SBA 2012). This is a broad definition for general advocacy and it does not qualify businesses for programs and services.

At the state and local levels, governments may adopt federal standards and/or modify or create separate standards when defining small businesses for their own policy and programs. As a result, each level of government may have its own definitions and metrics, which sometimes conflict. This section does not present a broad comparison between U.S. states and municipalities. It presents examples that show the contrast. The intent is not to review hundreds of examples, but simply to highlight the complexity. Specific examples from the state of Maryland and one local government from the same state illustrate the issue.

The state of Maryland recently codified its definition of a small business through legislation for the purposes of state procurement. The excerpt below describes the first portion of Maryland's legal definition of a small business.

> "Small business" is a business, other than a broker, that meets the following criteria:
>
> - Independently owned and operated;
> - Is not a subsidiary of another business; and
> - Not dominant in its field of operation.
>   (Maryland State Finance and Procurement Code 14-501)

The second set of criteria use different metrics across six sectors (wholesale, retail, manufacturing, service, construction and architectural and engineering). The metrics differ by sector across the number of employees and gross sales (average of past three fiscal years). As shown in Table 6.1, the law identifies six operational sectors (wholesale, retail, manufacturing, service, construction, architectural and engineering) and uses three employee size standards (25, 50 and 100) and six gross sales figures ($2M, $3M, $4M, $4.5M, $7M and $10M).

In contrast, a local example from Montgomery County, Maryland, shows another set of criteria for the definition of small business. The Montgomery County Local Small Business Reserve (LSBR) program reserves 20 percent of county agency purchases for small businesses. The basic screening criteria include that the business: 1) has its principal place of business in the county; 2) is independently owned and operated; 3) is not a subsidiary of another business; 4) generates a significant amount of economic activity in the county; and 5) complies with state prevailing and local living wage provisions. Table 6.2 shows additional program eligibility criteria. The criteria include five different business types (retail, wholesale,

Table 6.1 Maryland Small Business Size Criteria

Sector	Number of Employees	Gross Sales (in millions)
Wholesale	Not more than 50 persons	$4
Retail	Not more than 25 persons	$3
Manufacturing	Not more than 100 persons	$2
Service	Not more than 100 persons	$10
Construction	Not more than 50 persons	$7
Architectural and Engineering	Not more than 100 persons	$4.5

Source: Constructed from elements as found in the State Finance and Procurement, Maryland Code Annotated (2004) § 14-501

Table 6.2 LSBR Program Size and Sales Criteria for Montgomery County, Maryland

Business Type	Employee Limit		Prior 3 Years' Average Sales
Retail	30	or	$5,000,000
Wholesale	30	or	$5,000,000
Service	50	or	$5,000,000
Construction	50	or	$14,000,00
Manufacture	40	or	$14,000,00

Source: Constructed from elements as found from Montgomery County, State of Maryland, *Local Small Business Reserve*, 2015.
www.montgomerycountymd.gov/PRO-LBP/Eligibility.html

service, construction, manufacturing), each with three employee size limits (30, 40 and 50) and two average sales figures ($5M and $14M).

The state and local policy examples from Maryland show several differences. The policies differ in identified sectors (six by state and five by local government). Employee limits also differ among the state and local programs. Most interesting is the retail sector identified by both programs. The state example limits employees to twenty-five, while the local program allows for thirty employees. Differences in sales limits are also significant, with the largest disparity in the manufacturing sector ($10M difference in limits). The largest difference in the two programs lies in the qualifying term: "and" at the state level and "or" at the local level. The state program requires meeting both sales and employee limits, whereas the local program indicates that the business must meet either the employee limit or sales limits.

The federal, state and local examples show that the definition of a small business is not a static concept, even in law. Four additional issues further complicate the small-business policy landscape across levels of government. These include non-employer businesses, economic and social disadvantage, gender and geographic location.

The first issue is the non-employer business. A non-employer business "has no paid employees, has annual business receipts of $1,000 or more ($1 or more in the construction industries), and is subject to federal income taxes" (U.S. Census Bureau n.d.). Individuals who operate these unincorporated businesses are self-employed and this may or may not be their primary source of personal income. Three quarters of all U.S. businesses are non-employer businesses, but they account for only 3 percent of business revenue (Acs *et al.* 2009). These businesses are on Main Streets across America and may be located between small businesses and large corporate entities. In almost all cases, the difference is not discernible to the casual observer.

The second complicating issue shows in federal, state and local government programs targeted at assisting socially or economically disadvantaged small businesses. Here programs define a subset of small business based upon the demographics of individuals that own or control the small business. Among several programs are the Minority Small Business and Capital Ownership Development Program (known as the "8(a) program"); the Small Disadvantaged Business program and the Women-Owned Small Business Federal Contract Program, as well as the Department of Transportation Disadvantaged Business Enterprises program. Programs have slight differences in definitional elements. Here, the SBA 8(a) program serves as the example for comparison.

The 8(a) program requires that a small business be "unconditionally owned and controlled by one or more socially and economically disadvantaged individuals who are of good character and citizens of the

United States" (13 CFR 124.101). The program is codified in federal law-provided specific definitions and criteria for socially and economically disadvantaged status. The law defines socially disadvantaged individuals as

> those who have been subjected to racial or ethnic prejudice or cultural bias within American society because of their identities as members of groups and without regard to their individual qualities. The social disadvantage must stem from circumstances beyond their control.
> (13 CFR 124.103)

It further identifies specific groups presumed to be socially disadvantaged. Individuals from the following group are socially disadvantaged under the law:

> Black Americans; Hispanic Americans; Native Americans (Alaska Natives, Native Hawaiians, or enrolled members of a Federally or State recognized Indian Tribe); Asian Pacific Americans (persons with origins from Burma, Thailand, Malaysia, Indonesia, Singapore, Brunei, Japan, China (including Hong Kong), Taiwan, Laos, Cambodia (Kampuchea), Vietnam, Korea, The Philippines, U.S. Trust Territory of the Pacific Islands (Republic of Palau), Republic of the Marshall Islands, Federated States of Micronesia, the Commonwealth of the Northern Mariana Islands, Guam, Samoa, Macao, Fiji, Tonga, Kiribati, Tuvalu, or Nauru); Subcontinent Asian Americans (persons with origins from India, Pakistan, Bangladesh, Sri Lanka, Bhutan, the Maldives Islands or Nepal); and members of other groups designated from time to time by SBA according to procedures set forth at paragraph (d) of this section. Being born in a country does not, by itself, suffice to make the birth country an individual's country of origin for purposes of being included within a designated group.
> (13 CFR 124.103)

The law also details the process to petition the SBA for inclusion of additional groups as socially disadvantaged.

The 8(a) program requires a status as both socially and economically disadvantaged. The law defines economic disadvantage in terms of "socially disadvantaged individuals whose ability to compete in the free enterprise system has been impaired due to diminished capital and credit opportunities as compared to others in the same or similar line of business who are not socially disadvantaged" (13 CFR 124.104). The law sets earning limits at $250,000 per year and total personal assets of less than $4 million (13 CFR 124.104) for individuals that own or control the small business.

As described above, the 8(a) program is gender-neutral. Gender is the third complicating issue for defining small businesses. The federal government also sponsors programs targeted at gender. The Women-Owned Small Businesses (WOSB) and Economically Disadvantaged Women Owned Small Businesses (EDWOSB) programs provide preferences for federal contracts for small businesses owned by women. The program defines a WOSB as "not less than 51 percent unconditionally and directly owned and controlled by one or more women who are United States citizens" (13 CFR 127.200). For an EDWOSB, the qualifier is added that the women owner(s) are also "economically disadvantaged" (13 CFR 127.200). In our example state, Maryland uses the core elements of several federal programs combined into a state-level program that includes small businesses controlled by women and/or individuals who are socially and economically disadvantaged. The Maryland Minority Business Enterprise (MBE) Program provides for preferential contracting for these small businesses as certified by the state (Maryland State Finance and Procurement Code Ann. § 14-301).

The fourth complicating issue for defining small business is geography. The Historically Underutilized Business Zones (HUBZone) SBA program identifies geography as a factor in defining small business. The goal of the HUBZone program is "to promote economic development and employment growth in distressed areas by providing access to more federal contracting opportunities" (SBA n.d.) The SBA defines areas as HUBZones by census tract. The program outlines the following criteria for eligibility:

- It must be a small business by SBA standards;
- It must be owned and controlled at least 51 percent by U.S. citizens, or a Community Development Corporation, an agricultural cooperative or an Indian tribe;
- Its principal office must be located within a "Historically Underutilized Business Zone," which includes lands considered "Indian Country" and military facilities closed by the Base Realignment and Closure Act; and
- At least 35 percent of its employees must reside in a HUBZone.

(SBA n.d.)

The multitude of federal, state and local definitions and metrics and the vast differences in the few examples from three levels of government in affecting a single state show the differentiation and complexity involved. When multiplied across the approximate 87,000 jurisdictions in the U.S., the problem grows exponentially. In short, from a policy perspective, the answer to the question *what constitutes a small business* is … it's complicated.

The review of government policy shows small business defined by sector, number of employees and total receipts (and in some cases other specific sector metrics) and considers the individual demographics of ownership to include socioeconomic status, gender and the geographic location of the business. The nature of these policies and programs indicates that businesses which meet these criteria require additional programs and opportunities during times of normal operation, absent the conditions brought about by disaster. If these businesses require special government programs during normal operation, the logic follows that these businesses are also in need of additional assistance during disasters.

Small Business in Emergency Management Research: Definitions and Research Findings

This section focuses on small business research across the phases of disaster and the metrics used to define a business as a small business. Tierney (2006) observes that limited emergency management research exists on business as a unit of analysis. More recent interest in business continuity and potential impacts of climate change has renewed emergency-management research interest in businesses. The following paragraphs first review the definitions of small business used by researchers, then the discussion shifts to examine research findings concerning businesses across each of the four phases of disaster: preparedness, response, recovery, mitigation.

As the following review shows, without common definitions for the inclusion or exclusion of cases for study, knowledge of businesses in disaster remains a patchwork at best. Researchers examining businesses take one of two approaches to defining small businesses in their studies. The first approach is to provide a specific definition of a small business. These definitions rely exclusively on the number of employees to define business size. The limited definitions either use government policy as a basis or are researcher constructed based on data. Many of the data definitions develop out of simple analytical convenience. Other researchers provide no rationale for their definition. In the second approach, the researcher does not provide a specific definition of small business, but uses number of employees as either a continuous variable or an ordinal variable grouped in size categories.

The Disaster Research Center (DRC) at the University of Delaware conducted the largest group of studies on businesses in disaster. These studies define small and large business in terms of the number of employees. Several of the studies define a small business as one with less than twenty employees, whereas large business had more than twenty (Dahlhamer & D'Souza 1995; 1997; Tierney, Nigg, & Dahlhamer 1996; Tierney 1997). The other major empirical DRC study by Webb, Tierney, & Dahlhamer (2002) uses ordinal categories to group businesses (1–4, 5–9, 10–14, 15–29, 30 or more) and does not specifically define those of a certain size as small businesses.

Generally, the point of demarcation in these and other studies is not empirically explained other than as a matter of sampling and analytical convenience. A footnote in the study by Dahlhamer & D'Souza (1995) provides the only explanation. The note states that a larger definition would result in "too few large businesses being selected for the study" (Dahlhamer & D'Souza 1995, p. 10). In another recent example, Corey & Deitch (2011) set the definition of small business at less than fifty employees without explanation. Howe (2011) divides business size into five categories (less than 5, 5–19, 20–99, 100–499, greater than 500). Similarly, Yoshida & Deyle (2005) divide into five size categories (0–4, 5–9, 10–19, 20–49, 50–99). Three studies (Sadiq & Weible 2010; Sadiq 2010; 2011) use seven categories (1–9, 10–19, 20–49, 50–99, 100–249, 250–499, >500) and define small organizations as those with less than nineteen employees. Across these and other studies, researchers do not explain the rationale for their ordinal divisions.

Table 6.3 captures the specific definitions used in studies examining small business. The table shows the nine different definitions used among nineteen studies containing definitions of small business. Despite sampling and analytical convenience, at what size do differences appear in management, decision-making and other factors critical to emergency management? This is a key question for research.

Finally, studies by Alesch *et al.* (2001), Danes *et al.* (2009) and Haynes, Danes, & Stafford (2011) further complicate the small business distinction. The study by Alesch *et al.* (2001) combines small business and non-profit

Table 6.3 Definitions of Small Business in Emergency Management Research

Definition	Study
500 or fewer employees	Haynes, Danes, & Stafford 2011; McDonald, Florax & Marshall 2014
200 or fewer employees	Schrank *et al.* 2013; Marshall *et al.* 2015
Fewer than 100 employees	Alesch *et al.* 2001; Yoshida & Dele 2003; Flynn 2003; Runyan 2006; Howe 2011
Fewer than 50 employees	Corey & Deitch 2011
Fewer than 20 employees	Dahlhamer & D'Souza 1995; Tierney *et al.* 1996; Dahlhamer & D'Souza 1997; Tierney 1997; Dahlhamer & Tierney 1998
19 or fewer employees	Sadiq 2011
15 or fewer employees	Drabek 1991
10 or fewer employees	Landahl 2013
5 or fewer employees	Chang & Falit-Baiamonte 2002

entities together in a study of recovery. Should studies pair non-profits and small businesses? The studies by Haynes *et al.* (2011) and Danes *et al.* (2009) also advance the impact of disasters on a class of businesses defined as "family firms." How are family firms different from small business?

Emergency Management Research: Categorizing Small Business

Another conceptual issue complicating definitions and limiting the generalizability of research is categorization by type of business activity. The question for research is whether the type of business activity is significant for disaster outcomes. Earlier in this chapter, policy definitions of small business for access to programs at the federal level centered on metrics differentiated by North American Industry Classification System (NAICS) codes. The Office of Management and Budget developed the NAICS (adopted in 1997) with North American economic partners. The NAICS replaced the Standard Industrial Classification (SIC) code system for many government programs; however, the U.S. Securities and Exchange Commission continues to use the SIC system. The state and local policy definition examples from Maryland showed additional differences in categorization for policy. Those examples show different categories and descriptions for business activities even among jurisdictions within a single state.

In research, a similar problem develops that limits the applicability of findings. The DRC studies introduced above use SIC codes. These studies use five sectors to divide business activities: 1) wholesale and retail; 2) manufacturing, construction and contracting; 3) business and professional services; 4) finance, insurance and real estate; 5) other (Dahlhamer & D'Souza 1995; Tierney *et al.*, 1996; Dahlhamer & D'Souza 1997; Tierney 1997; Dahlhamer & Tierney 1998; Webb *et al.* 2003). Other studies use SIC codes or NAICS codes, or develop their own categories for the division of business activities. For example, Chang & Falit-Baiamonte (2002) use three categories of business activity: 1) retail; 2) manufacturing; 3) producer services. Dietch & Corey (2011) also use three categories but differ from Chang & Falit-Baiamonte, instead choosing to focus on: 1) construction/manufacturing; 2) retail and wholesale trade; 3) the service industry. A few studies focus on a single specific sector, and other research is developed in only two specific sectors: 1) tourism (Drabek 1991; 1994; 1995); and 2) the chemical industry (Quarantelli *et al.* 1979; Lindell & Perry 1998). As the review of research in the phases of disaster shows, sector matters. The lack of unity in research results in unanswered questions concerning how sectors are defined and business activities categorized. This results in significant gaps in research-based understanding.

Small Business Definitions in Policy and Research

This review of definitional issues in policy and research shows that small business is not a static concept. The differences in definitions for policy across levels of government show the complexity of the issue. Definitions of small business in research also lack consistency. The review shows eight separate definitions within seventeen studies that provide specific definitional criteria. What constitutes a small business is obviously a matter of contention in both research and policy. For research, the conceptualization and operationalization problem hinders the spiral development of knowledge of business in disaster. This lack of clarity continues to cloud knowledge of small businesses for application to emergency-management policy and practice. The review of conceptual issues creates many more questions than answers. The lack of specificity leads to non-replicable studies, resulting in the continuation of the previously described patchwork of knowledge.

Businesses in Disaster: Research across the Phases of Emergency Management

Emergency management research generally focuses on one or more of the four phases of emergency management: mitigation, preparedness, response and recovery. The phased approach has been in use for more than three decades, since the adoption of comprehensive emergency management (NGA 1979). The phases are not mutually exclusive (Neal 1997). Research often examines issues within more than one phase, as the lines between these phases are not clearly discernable. The intent here is not to enter conceptual debates regarding each phase of emergency management, but to use them as a general framework to organize our review of existing emergency management research with businesses as the unit of analysis.

This review is organized around research that measures outcomes related to each of the four phases. Where possible, discussion includes construction of related dependent variables and associated factors. Many of the studies appear in multiple sections, as findings apply to more than a single phase. The discussion begins with the recovery phase, as it has been the subject of the largest group of studies. The review concludes with discussion of the extremely limited research regarding the response phase.

Recovery

The largest body of research on business as a unit of analysis focuses on businesses' recovery from disaster. The recovery process also struggles with its own conceptual issues (Quarantelli 1998; 1999). Researchers generally conceptualize and measure disaster recovery as self-reported perceptions of business owners (Dahlhamer & Tierney 1996; Webb, Tierney, & Dahlhamer 2000; Webb, Tierney, & Dahlhamer 2000; 2002). Dahlhamer & Tierney (1996) measured business recovery based on self-reporting by business across a trichotomous dependent variable. Businesses self-reported they were "worse off, about the same, or better off" following the relevant disaster (Dahlhamer & Tierney 1996). Webb et al. (2002) constructed the dependent variable as a composite of four self-reported items: number of employees, number of customers or clients served, business profits and whether the business was "worse off, about the same, or better off" following the disaster.

These and other studies identify several factors related to business recovery. Table 6.4 identifies factors linked in more than twenty studies targeted at business recovery. It modifies and expands upon an initial similar summary presented by Webb et al. (2002). This iteration updates and modifies the original table through removal of non-disaster research findings and replacement with specific disaster-related research from the past thirteen years. The groundbreaking research of Webb et al. (2002) required the use of non-disaster research to develop variables within the study, but research has advanced such that targeted disaster research can now support identification of related factors. Table 6.4 includes several factors identified in disaster research from the original table, with consolidation of similar concepts across studies where possible.

The review of studies in Table 6.4 identifies twenty-four factors associated with business recovery from disaster. The studies frequently cite nine factors: 1) business size; 2) impact on customers; 3) building damage; 5) business sector; 6) lifeline services; 6) female ownership; 7) impact on employees; 8) larger economic trends; and 9) disruption of operations.

Preparedness

The conceptualization of preparedness also faces a number of issues. Preparedness measures generally include a set of elements consisting of plans, supplies and actions. This often presents in the form of a checklist of items and actions. Generally, researchers operationalize a preparedness dependent variable as a composite measure derived from the number of checklist elements which a household or business possesses or engages. These often combine into a set number of items organized into ordinal categories of high, moderate or low preparedness. For example, Tierney (1997) presents a fifteen-item list of business-preparedness elements that forms the basis for several of the studies that follow. Some studies develop slightly different checklists for specific hazards (such as earthquakes).

Some of the aforementioned recovery literature appears in this section. Many of these studies are both successful and unsuccessful at linking

Table 6.4 Factors Associated with Business Recovery

Factor	Source
Firm Age	Webb *et al.* 2002; Marshall *et al.* 2015
Own or Lease Business Site	Durkin 1984
Business Size	Kroll *et al.* 1991; Tierney 1997; Chang & Falit-Baiamonte 2002; Zolin & Kropp 2007; Zhang *et al.* 2009
Impact on Customers	Tierney 1997; Alesch *et al.* 2001; Zhang *et al.* 2009; Deitch & Corey 2011; Corey & Deitch 2011
Product Competition	Alesch *et al.* 2001; Runyan 2006; Chang 2010
Product Necessity	Alesch *et al.* 2001
Place on Industry Curve	Alesch *et al.* 2001
Stability Prior to Disaster	Durkin 1984; Alesch *et al.* 2001; Webb *et al.* 2002
Entrepreneurial Skill	Alesch *et al.* 2001; Zolin & Kropp 2009
Building Damage	Durkin 1984; Alesch *et al.* 2001; Webb *et al.* 2002; Corey & Deitch 2011
Business Sector	Kroll *et al.* 1991; Losocco & Robinson 1991; Dahlhamer & D'Souza 1997; Dahlhamer & Tierney 1998; Tierney 1998; Webb *et al.* 2002; Chang & Falit-Baiamonte 2002; Runyan 2006; Corey & Deitch 2011; Marshall *et al.* 2015
Lifeline Services	Rose *et al.* 1987; Kroll *et al.* 1991; Tierney *et al.* 1996; Tierney 1997; Webb *et al.* 2000; Runyan 2006; Corey & Deitch 2011; Asgary *et al.* 2012
Impact on Employees (Labor Shortage)	Tierney 1997; Dahlhamer 1998; Zolin & Kropp 2007; Zhang *et al.* 2009; Corey & Deitch 2011; Deitch & Corey 2011
Larger Economic Trends	Tierney 1997; Dahlhamer 1998; Webb *et al.* 2000; Webb *et al.* 2002; Chang 2010; Deitch & Corey 2011
Female Ownership	Tiggues & Green 1994 (−); Danes *et al.* 2009 (+); Haynes *et al.* 2011 (+); McDonald *et al.* 2014 (−), Marshall *et al.* 2015 (−)
Minority Ownership	Marshall *et al.* 2015 (−)
Veteran Ownership	Marshall *et al.* 2015 (−)
Primary Market	Dahlhamer & Tierney 1998; Webb *et al.* 2002; Zhang *et al.* 2009
Disaster Experience	Drabek 1994; Marshall *et al.* 2015
Duration of Closure	Alesch *et al.* 1993; Webb *et al.* 2002
Disruption of Operations	Durkin 1984; Tierney 1997; Webb *et al.* 2002; Corey & Deitch 2011
Preparedness	Dahlhamer & D'Souza 1997; Chang & Filat-Baiamonte 2002; Runyan 2006
Disaster Assistance	Runyan 2006; Haynes *et al.* 2011
Cash Flow Issues	Runyan 2006; Zolin & Kropp 2007; Marshall *et al.* 2015

business pre-event preparedness to recovery in the aftermath. Although two studies in Table 6.4 identify preparedness as significant predictors of recovery, many more did not find a relationship. Several explain that a prepared business may face many factors beyond its control—such as lifelines, impact on customer base and employees, and so on—that cannot be affected by individual business-preparedness (Alesch *et al.* 2001; Webb *et al.* 2002; Tierney 2006).

Business preparedness is the subject of fewer studies than business recovery. The studies in existence generally find limited preparedness measures are undertaken by businesses (Drabek 1994; Tierney 1997; Webb *et al.* 2000; Runyan 2006; Landahl 2013). Table 6.5 shows the factors linked to preparedness from nine studies. The table identifies ten variables associated with business disaster-preparedness. Studies cite three independent variables: business size; prior experience with disaster; and sector of the business frequently associated (four or more studies) with business preparedness.

Mitigation

Similar to business recovery and preparedness, conceptual issues also abound for mitigation. Limited research also inhibits research-based understanding and conceptual development. The few studies lack cohesion

Table 6.5 Factors Associated with Business Preparedness

Factor	Source
Business Size	Quarantelli *et al.* 1979; Drabek 1991; 1994; Dahlhamer & D'Souza 1997; Howe 2011; Landahl 2013
Own or Lease Location	Dahlhamer & D'Souza 1997; Howe 2011
Prior Experience with Disaster	Mileti *et al.* 1993; Banerjee & Gillespie 1994; Drabek 1994; Dahlhamer & D'Souza 1997; Flynn 2009; Mayer *et al.* 2009; Howe 2011
Business Sector	Mileti *et al.* 1993; Drabek 1991; 1995; Dahlhamer & D'Souza 1997; Howe 2011
Age of Business	Quarantelli *et al.* 1979 (−); Drabek 1991; (+)
Multiple Business Locations	Drabek 1991; 1995
Location of Primary Market	Howe 2011
Exposure to Storm Surge	Howe 2011
High-Risk Perception	Howe 2011
Seek Risk Information	Howe 2011; Mileti *et al.* 1993
Reduced Physical Damage	Xiao & Peacock 2014
Access to Expertise	Landahl 2013

in defining mitigation activities and often comingle activities from other phases. For example, in studies of organizations in the Memphis, TN area (Sadiq & Weible 2010; Sadiq 2010; 2011), a list of ten risk-reducing activities is provided and organized in terms of low- and high-effort activities. This construct is similar to the preparedness checklist; however, these studies present the ten elements as dependent variables. These studies also define mitigation in terms of an organizational activity (Sadiq & Weible 2010; Sadiq 2010; 2011). For their purposes, organizations include both public and private-sector entities. This limits the specific applicability of conclusions to businesses alone, raising the issue of business as a unique organizational type.

Other studies define mitigation in terms of hazard adjustments. Montes de Oca (2011) defines mitigation in terms of adaptive measures and coping actions. Lindell and Perry (1998) describe these in terms of three elements: hazard assessment, hazard mitigation and emergency preparedness actions. Yoshida and Deyle (2005) operationalize mitigation in terms of three elements: a business continuity plan, one or more structural mitigation activities and purchase of one or more types of insurance. Sadiq and Weible (2010) bifurcate ten identified mitigation activities into high- and low-effort. Sadiq (2011) groups these same activities into active and passive mitigation. Overall, these studies show a lack of cohesion and spiral development in the study of businesses' mitigation activities.

Despite these major differences, Table 6.6 shows the factors linked to mitigation from six identified studies. The table identifies five factors associated with mitigation activities by businesses. The variables include business size, business sector, risk perception, prior experience with disaster damage and access to expertise.

Response

The response section of this paper does not begin with the term "dependent variable." This is due to three issues. First, response consists

Table 6.6 Factors Associated with Business Mitigation Activities

Factor	Source
Access to Expertise	Yoshida & Deyle 2005
Business Size	Sadiq 2010; Sadiq & Weible 2010; Sadiq 2011
Business Sector	Yoshida & Deyle 2005
Risk Perception	Yoshida & Deyle 2005; Sadiq 2010
Experienced Direct Damage	Lindell & Perry 1998; Montes de Oca 2011
Reduced Physical Damage	Xiao & Peacock 2014

of many sub-actions that are not additive. As such, it cannot be operationalized in the same manner as preparedness and mitigation variables. Second, response is dynamic and subject to many intervening variables. Finally, business as a unit of analysis in response is almost completely devoid of empirical study. Other works on the topic of response are theoretical and are discussed in the final section of this chapter.

Drabek (1995) presents the only empirical study of businesses as a unit of analysis in response. The study focuses on evacuations within the tourism industry. The study is sector-specific and specific to one response activity. The study of response actions requires this type of specificity due to the complexity of actions involved. Drabek (1995) studied 185 firms in six communities recently impacted by disasters to determine elements significant to evacuation responses. He identified that managers that engage in "more rapid and elaborate evacuation behavior had done more extensive planning [and] were responsible for more complex and larger lodging firms that had been evacuated previously" (Drabek 1995, p. 13). These findings dovetail with research in other phases that identify business size, business sector and previous experience as variables of significance. Unfortunately, the study of response is limited to one study in one specific industry.

Emergency Management Research Summary

The review of study findings reinforces that emergency management research presents islands of knowledge about businesses as a unit of analysis. The discussion shows that conceptualization and operationalization of business size and sector is problematic for research. The review also evidences the scarcity of research in both mitigation and response phases and limited development and fragmentation due to definitional difficulties.

The revival of business-related research in the post-Katrina environment also presents questions of applicability. The problem of disaster scale plagues all research in the field of emergency management. It raises the question of whether knowledge of events characterized as either a "disaster" or a "catastrophe" is transferrable among each other (Quarantelli 2008; Tierney 2006).

The review of research shows commonality in several variables of significance across research in the phases of emergency management. The two key common variables are business size and business sector. According to the studies, these variables influence actions in mitigation, preparedness, (limited) response and recovery.

Many researchers generally view business size as a proxy for resources (Tierney, 2006; Zhang et al., 2009). It makes sense, and research shows that those with more resources are more likely to engage in preparedness and mitigation measures and more likely to recover following a disaster. The sector of business operation also influences activities across the phases. The Disaster Research Center (DRC) studies found the financial, insurance and real estate (FIRE) sector was more likely to prepare and more likely to recover. Drabek (1995) found that lodging firms in the tourism industry are more likely to prepare and respond quickly for potential evacuations. Sector differences could result from factors such as industry self-imposed standards or government regulation. In addition to the need for consistency across studies, research needs to explain if differences exist across business size and determine what size limits are significant.

Small Businesses as a Vulnerable Population

According to Cutter, Boruff and Shirley (2003), researchers cast vulnerability in one of three contexts. The contexts include: 1) an exposure model that identifies conditions that make people or places vulnerable; 2) vulnerability as a function of social conditions; or 3) measurement of specific places for potential exposures and ability to resist (Cutter *et al.* 2003). Chang and Falit-Baiamonte (2002) first advanced the discussion linking the concept of social vulnerability as a function of social conditions to businesses. The synthesis of research in this chapter provides the opportunity to reengage with and expand the discussion initiated by Chang and Falit-Baiamonte (2002).

Blaikie *et al.* (1994) define vulnerability as "the characteristics of a person or group in terms of their capacity to anticipate, cope with, resist and recover from the impact of natural hazards" (p. 9). The definition includes both individual and group characteristics. Social vulnerability theory contends that disproportionate consequences of disaster are the result of societal conditions that place individuals with certain characteristics at a disadvantage. Essentially, disasters have uneven effects on those who face equal exposure to hazards. Cutter *et al.* (2003) identify the generally accepted characteristics that influence social vulnerability as: socioeconomic status (income, political power and prestige); gender, race and ethnicity; and age (the elderly and young children).

The policy review shows that government programs specifically target businesses based on size, geographic location and demographics of business ownership, as they are vulnerable to economic conditions absent disaster. The intent is not to review economic research in normal operating conditions, but to identify that research findings support the necessity of these programs. For example, research in non-disaster conditions shows disparities in success rates and income for businesses based on size (e.g. Aldrich & Auster 1986), ownership by women (e.g. Loscocco *et al.* 1991; Tiggues & Green 1994) and ownership by racial minorities (e.g. Fairlie & Robb 2008). This research and the existence of government programs support the assertion that these businesses are vulnerable based upon size and the gender and racial composition of ownership.

Smaller businesses are less likely to recover after disasters (Kroll *et al.* 1991; Tierney 1997; Chang & Falit-Baiamonte 2002; Zolin & Kropp 2007; Zhang *et al.* 2009). The review of disaster research shows that the aforementioned vulnerabilities carry into disaster mitigation, preparedness and recovery, with a lack of the resources necessary to prepare for and/or cope with the effects of disaster. The research shows that smaller businesses are less likely to recover after disaster (Kroll *et al.* 1991; Tierney 1997; Chang & Falit-Baiamonte 2002; Zolin & Kropp 2007; Zhang *et al.* 2009). Smaller businesses are also less likely to undertake preparedness measures (Quarantelli *et al.* 1979; Drabek 1991; 1994; Dahlhamer & D'Souza 1997; Howe 2011; Landahl 2013) and/or mitigation measures (Sadiq 2010; Sadiq & Weible 2010; Sadiq 2011). Size is a general proxy for resources (Tierney 2006; Zhang *et al.* 2009). These businesses, similarly to individuals considered socially vulnerable, lack resources to prepare for and/or cope with the effects of disaster.

Outcomes also differentiate by business sector (Kroll *et al.* 1991; Losocco & Robinson 1991; Dahlhamer & D'Souza 1997; Dahlhamer & Tierney 1998; Tierney 1997; Webb *et al.* 2002; Chang & Falit-Baiamonte 2002; Runyan 2006; Corey & Deitch 2011; Marshall *et al.* 2015), preparedness (Mileti *et al.* 1993; Drabek 1991; 1995; Dahlhamer & D'Souza 1997; Howe 2011) and mitigation (Yoshida & Deyle 2005). The research suggests that in general, businesses in the finance, insurance and real estate sectors (defined in DRC studies) are more likely to engage preparedness and mitigation activities and more likely to recover.

Conclusion: Implications for Research, Policy and Practice

Societal risks regarding disasters are increasing (World Economic Forum, 2015). Each year, the U.S. faces a series of natural and man-made disasters that cause hundreds of deaths and cost billions of dollars. Even if the probability or intensity of such risks remains constant, population growth, alongside economic, infrastructural, and technological development, will inherently result in a concomitant increase in places prone to disasters (Mileti 1999). Small businesses, as critical components of society and economies, remain vulnerable. The research review shows that small businesses are disproportionately vulnerable and face disproportionate impacts from disaster. This chapter advances the proposition to characterize small businesses as a vulnerable population. From an emergency management research, policy and practice perspective, the question follows: what do we do about it? This final section examines the implications of the elements reviewed for future research, emergency management policy and the practice of emergency management.

Implications for Research

This chapter shows there are two main issues for future research. The first relates to conceptualization and operationalization of terms for research. Tierney (2006) captures the key theme of this review and frames the problem for the study of businesses as a unit of analysis. Addressing the question "do businesses recover," Tierney (2006) notes: "the answer to this question depends how in part the question is asked: how recovery is conceptualized, how the concept is operationalized, what types of businesses are selected for study, and how studies are conducted" (p. 285). Although her comment targets business recovery, the review of the literature shows applicability beyond this. Key issues in research include conceptualizing and operationalizing dependent variables and key business characteristics. The review shows issues in defining business characteristics, such as the many meanings of "small business." The unmet research goal is to determine if any of this matters, and particularly the business sizes at which differentiations occur. As the review shows, we take the ill-defined term of "small business" and marry it with the just as ill-defined term "disaster." These issues hamper the spiral development of knowledge about businesses in disaster.

The second issue is theoretical. Tierney (2006) cites a lack of application of organizational and general social science theories to research on businesses. Of the studies in this review, few cite theoretical ties. The works by Drabek (1991; 1994; 1995) cite the stress-strain framework for organizational analysis (Haas & Drabek 1973). Few, if any, others cite theoretical bases. Tierney (2006) advocates the application of normal accident theory (Perrow 1984), high reliability theory (Weick *et al.* 1999) and crisis-management theories. Birkland and Nath (2000) argue that business theories, such as those used in the crisis-management literature, underestimate the political dimensions of disaster. In both cases, much of the research focusing on business is devoid of a theoretical framework.

The conceptualization and operationalization issues related to small business combine with the absence of theory to inhibit the development of knowledge for businesses in disasters. The importance of the topic is without question. Future research must reconcile these issues to build consistent knowledge on this critical topic.

Implications for Policy

The policy review in this chapter shows complexity and disparity in legal definitions of small business. The examples show the depth of the disparity even within a single state. When compounded across a potential 87,000 jurisdictions in the U.S., the problem expands exponentially. While interesting, the legal definition of small business is not of real consequence for practice. It is a function of American federalism and the purview of state and local governments to define the parameters of individual programs. Small business's vulnerability to disaster is a local problem—a problem that is more in need of support from substantive emergency-management research than of federal policy interventions. From a policy perspective, governments may consider tax incentives for the adoption of preparedness and mitigation measures (Dahlhamer & D'Souza 1995; Sadiq 2010). As a condition of accessing small-business programs, governments could require certain preparedness and mitigation activities to be undertaken. This would serve to reinforce emergency management outcomes within the framework of economic programs.

Implications for Practice

Local emergency management is the proverbial "tip of the spear" for engaging small businesses as a vulnerable population. Emergency managers can take several approaches within the current policy framework to increase mitigation and preparedness activities among this population. The first is to ensure that emergency managers and first-response organizations understand the needs of businesses (Zolin & Kropp 2007). Local emergency planners must recognize the importance of lifeline services such as power, water and transportation, to ensure the needs of business are considered in planning. In turn, emergency managers need to help small-business owners foster an understanding of the community risk profile in order to recover more quickly (Webb et al. 2000). Understanding the community risk profile can allow small-business owners to better assess current insurance needs and find gaps in coverage prior to events (Alesch et al. 2001; Landahl 2013). Yoshida and Deyle (2005) identify that hazard-education programs can show business owners the "benefits of hazard insurance, business continuity plans, and specific structural mitigation options" (p. 7), thereby increasing community preparedness. Local emergency managers provide the access to preparedness and mitigation expertise that small businesses lack (Yoshida & Deyle 2005; Landahl 2013).

Finally, continuous outreach efforts by local emergency-management agencies are critically important. The research findings show that small businesses in certain sectors are less likely to engage preparedness and mitigation measures. With equally limited government resources, efforts by emergency managers should specifically target these businesses (Webb et al. 2000). In addition, emergency managers should leverage public–private partnerships for preparedness (Abou-Bakr 2013). Existing organizational and social networks provide a method for reaching businesses (Landahl 2013). Engagement and partnerships with community business organizations such as chambers of commerce and other economic partnership groups (such as downtown partnerships) provide a vehicle for access to businesses to carry forward preparedness and mitigation messages (Yoshida & Deyle 2005; Abou-Bakr 2013). In summary, as Webb *et al.* (2000, p. 90) describe:

> the more businesses work with governmental preparedness organizations in the communities in which they operate to reduce potential communitywide disaster impacts and streamline the recovery process, the more confident they can be that their own disaster-related problems will be less severe.

Bibliography

Abou-Bakr, A.J. *Managing Disasters Through Public–Private Partnerships*. Washington, DC: Georgetown University Press, 2013.

Acs, Z.J., Headd, B. & Agwara, H. "Nonemployer Start-up Puzzle." U.S. Small Business Administration, 2009. Accessed August 7, 2016. www.sba.gov/sites/default/files/Nonemployer%20Start-up%20Puzzle.pdf

Aldrich, H. & E.R. Auster. "Even Dwarfs Started Small: Liabilities of Age and Size and their Strategic Implications." In *Research in Organizational Behavior*, edited by B.M. Staw & L.L. Cummings, 165–198. Greenwich, CT: JAI Press, 1986.

Alesch, D.J., Holly, J.N., Minter, E. & Nagy, R.A. *Organizations at Risk: What Happens When Small Businesses and Not-For-Profits Encounter Natural Disasters*. Fairfax, VA: PERI, 2001.

Alesch, D.J., Taylor, C., Ghanty, A.S. & Nagy, R.A. "Earthquake Risk Reduction and Small Business." In *Socioeconomic Impacts*, edited by K. J. Tierney & J. M. Nigg, 133–160. Monograph prepared for the 1993 National Earthquake Conference. Memphis, TN: Central United States Earthquake Consortium, 1993.

Asgary, A., Anjum, M.I. & Azimi, N. "Disaster Recovery and Business Continuity after the 2012 Flood in Pakistan: Case of Small Business." *International Journal of Disaster Risk Reduction* 2 (2012): 46–56.

Banerjee, M.M. & Gillespie, D.F. "Strategy and Organizational Disaster Preparedness." *Disasters* 18 (1994): 344–354.

Birkland, T.A. & Nath, R. "Business and Political Dimensions in Disaster Management." *Journal of Public Policy* 20 (2000): 275–303.

Blaikie P., Cannon, T., Davis, I. & Wisner, B. *At Risk: Natural Hazards, People's Vulnerability, and Disaster*. London: Routledge, 1994.

Chang, S.E. "Urban Disaster Recovery: A Measurement Framework with Application to the 1995 Kobe Earthquake." *Disasters* 34 (2010): 303–327.

Chang, S.E. & Falit-Baiamonte, A. "Disaster Vulnerability of Businesses in the 2001 Nisqually Earthquake." *Environmental Hazards* 4 (2002): 59–71.

Corey, C.M. & Deitch, E.A. "Factors Affecting Business Recovery Immediately after Hurricane Katrina." *Journal of Contingencies and Crisis Management* 19 (2011): 169–181.

Cutter, S.L., Boruff, B.J. & Shirley, W.L. "Social Vulnerability to Environmental Hazards." *Social Science Quarterly* 84 (2003): 242–261.

Dahlhamer, J.M. "Rebounding from Environmental Jolts: Organizational and Ecological Factors Affecting Business Disaster Recovery." Doctoral Dissertation, Disaster Research Center: University of Delaware, 1998.

Dahlhamer, J.M. & D'Souza, M.J. "Determinants of Business Disaster Preparedness in Two U.S. Metropolitan Areas." Preliminary Paper #224, Disaster Research Center: University of Delaware, 1995.

Dahlhamer, J.M. & D'Souza, M.J. "Determinants of Business Disaster Preparedness in Two U.S. Metropolitan Areas." *International Journal of Mass Emergencies and Disasters* 15 (1997): 265–281.

Dahlhamer, J.M. & Tierney, K.J. "Winners and Losers: Predicting Business Disaster Recovery Outcomes Following the Northridge Earthquake." Preliminary Paper #243, Disaster Research Center: University of Delaware, 1996.

Dahlhamer, J.M. & Tierney, K.J. "Rebounding from Disruptive Events: Business Recovery Following the Northridge Earthquake." *Sociological Spectrum* 18 (1998): 121–141.

Danes, S.M., Lee, J., Amarapurkar, S. Stafford, K. Haynes, G. & Brewton, K.E. "Determinants of Family Business Resilience after a Natural Disaster by Gender of Business Owner." *Journal of Developmental Entrepreneurship* 14 (2009): 333–354.

Deitch, E.A. & Corey, C.M. "Predicting Long-Term Business Recovery Four Years after Hurricane Katrina." *Management Research Review* 34 (2011): 311–324.

Dilger, R.J. "Small Business Size Standards: A Historical Analysis of Contemporary Issues." Washington, DC: Congressional Research Service, 2012.

Drabek, T.E. "Disaster Responses within the Tourist Industry." *International Journal of Mass Emergencies and Disasters* 13 (1995): 7–23.

Drabek, T.E. "Disaster Evacuation and the Tourist Industry." Institute of Behavioral Science, University of Colorado, 1994.

Drabek, T.E. "Anticipating Organizational Evacuations: Disaster Planning by Managers of Tourist-Oriented Private Firms." *International Journal of Mass Emergencies and Disasters* 9 (1991): 219–245.

Durkin, M.E. "The Economic Recovery of Small Business after Earthquakes: The Coalinga Experience." Paper Presentation, International Conference on Natural Hazards Mitigation and Practice, New Delhi, India, 1984.

Fairlie, R.W. & Robb, A.M. *Race and Entrepreneurial Success: Black-, Asian-, and White-Owned Businesses in the United States.* Cambridge, MA: MIT Press, 2008.

Flynn, D.T. "The Impact of Disasters on Small Business Disaster Planning: A Case Study." *Disasters* 31 (2007): 508–515.

Friesma, H.P. *Aftermath: Communities after Natural Disaster.* Thousand Oaks, CA: Sage Publications, 1979.

Haas, J.E. & Drabek, T.E. *Complex Organizations: A Sociological Perspective.* New York: Macmillan, 1973.

Haynes, G.W., Danes, S.M. & Stafford, K. "Influence of Federal Disaster Assistance on Family Business Survival and Success." *Journal of Contingencies and Crisis Management* 19 (2011): 86–98.

Howe, P.D. "Hurricane Preparedness as Anticipatory Adaptation: A Case Study of Community Businesses." *Global Environmental Change* 21 (2011): 711–720.

Kroll, C.A., Landis, J.D., Shen, Q. & Stryker, S. "Economic Impacts of the Loma

Prieta Earthquake: A Focus on Small Business." Working Paper 91/187, University of California: Transportation Center and the Center for Real Estate and Economics, 1991.

Landahl, M.R. "Businesses and International Security Events: Case Study of the 2012 G8 Summit in Frederick County, Maryland." *Journal of Homeland Security and Emergency Management* 10 (2013): 609–629.

Lindell, M.K. & Perry, R.W. "Earthquake Impacts and Hazard Adjustment by Acutely Hazardous Material Facilities Following the Northridge Earthquake." *Earthquake Spectra* 14 (1998): 285–299.

Loscocco, K.A., Robinson, J., Hall, R.H. & Allen, J.K. "Gender and Small Business Success: An Inquiry into Women's Relative Disadvantage." *Social Forces* 70 (1991): 65–85.

Marshall, M.I., Niehm, L.S., Sydnor, S.B. & Schrank, H.L. "Predicting Small Business Demise after a Natural Disaster: An Analysis of Pre-Existing Conditions." *Natural Hazards* 79 (2015): 331–354.

Marshall, M.I. & Schrank, H.L. "Small Business Disaster Recovery: A Research Framework." *Natural Hazards* 72 (2014): 597–616.

Mayer, B.W., Moss, J. & Dale, K. "Disaster and Preparedness: Lessons from Hurricane Rita." *Journal of Contingencies and Crisis Management* 16 (2008): 14–23.

McDonald, T., Florax, R. & Marshall, M. "Informal and Formal Financial Resources and Small Business Resilience to Disasters." Paper Presentation, Agricultural and Applied Economics Association, 2014 AAEA and CAES Joint Annual Meeting, Minneapolis, MN, 2014.

Mileti, D. *Disasters by Design*. Washington, DC: Joseph Henry Press, 1999.

Mileti, D., Darlington, J., Fitzpatrick, C. & O'Brien, P. *Communicating Earthquake Risk: Societal Response to Revised Probabilities in the Bay Area*. Fort Collins, CO: Hazards Assessment Laboratory and Department of Sociology, Colorado State University, 1993.

Montes de Oca, P.H. "Past Disaster Damages as Drivers of Coping and Adaptive Strategies in Small and Medium Community Businesses." Paper Presentation, Belpasso International Summer School on Environmental and Resource Economics Belpasso, Italy, 2011.

Montgomery County, State of Maryland. *Local Small Business Reserve*, 2015. Accessed August 7, 2016. www.montgomerycountymd.gov/PRO-LBP/Eligibility.html

Neal, D.M. "Recasting the Phases of Disaster." *International Journal of Mass Emergencies and Disasters* 15 (1997): 239–264.

National Governors' Association (NGA). "Comprehensive Emergency Management: A Governor's Guide." National Governors' Association, Center for Policy Research, Washington, DC, 1979.

Perrow, C. *Normal Accidents: Living with High Risk Technologies*. New York: Basic Books, 1984.

Quarantelli, E.L. "Disaster Recovery: Research Based Observations on What It Means, Success and Failure, those Assisted and those Assisting." Preliminary Paper #263, Disaster Research Center: University of Delaware, 1998.

Quarantelli, E.L. "The Disaster Recovery Process: What We Know and Do Not Know from Research." Preliminary Paper #286, Disaster Research Center:

University of Delaware, 1999.

Quarantelli, E.L. "Conventional Beliefs and Counterintuitive Realities." *Social Research* 75, no. 3 (2008): 873–904.

Quarantelli, E.L., Lawrence, C., Tierney K. & Johnson, T. "Initial Findings from a Study of Socio-Behavioral Preparations and Planning for Acute Chemical Hazard Disasters." Disaster Research Center, Department of Sociology, The Ohio State University, 1979.

Rose, A., Benavides, J., Chang, S., Szczesniak, P., & Lim, D. "Regional Economic Impact of an Earthquake: Direct and Indirect Effects of Electricity Lifeline Disruptions." *Journal of Regional Science* 37 (1997): 437–458.

Rossi, P.H., Wright, J.D., Wright, S.R. & Weber-Burdin, E. "Are There Long-Term Effects of American Natural Disasters?" *Mass Emergencies* 3 (1978): 117–132.

Runyan, R.C. "Small Business in the Face of Crisis: Identifying Barriers to Recovery from Natural Disaster." *Journal of Contingencies and Crisis Management* 14 (2006): 12–26.

Sadiq, A. "Digging through Disaster Rubble In Search of the Determinants of Organizational Mitigation and Preparedness." *Risk, Hazards and Crisis in Public Policy* 1 (2010): 33–62.

Sadiq, A. "Adoption of Hazard Adjustments by Large and Small Organizations: Who Is Doing the Talking and Who Is Doing the Walking?" *Risk, Hazards and Crisis in Public Policy* 2 (2011): 1–17.

Sadiq, A. & Weible, V. "Obstacles and Disaster Risk Reduction: Survey of Memphis Organizations." *Natural Hazards Review* 11 (2010): 110–117.

Schrank, H.L, Marshall, M. Hall-Phillips, A.,Wiatt, R.F. & Jones, N.E. "Small-Business Demise and Recovery after Katrina: Rate of Survival and Demise." *Natural Hazards* 65 (2013): 2353–2374.

State Finance and Procurement, Maryland Code Annotated (2004) §14-501.

Tierney, K.J. "Business Impacts of the Northridge Earthquake." *Journal of Contingencies and Crisis Management* 5 (1997): 87–97.

Tierney, K.J. "Businesses and Disasters: Vulnerability, Impacts, and Recovery." In *Disaster Research Handbook*, edited by H. Rodríguez,. E.L. Quarantelli & R.R. Dynes, 275–296. New York: Springer, 2006.

Tierney K.J., Nigg, J.M., & Dahlhamer, J.M. "The Impact of the 1993 Midwest Floods: Business Vulnerability and Disruption in Des Moines." In *Disaster Management in the U.S. and Canada*, edited by R. Sylves and W. Waugh, 214–233. West Springfield, IL: Charles C. Thomas, 1996.

Tiggues, L.M. & Green, G.P. "Small Business Success among Men and Women Owned Firms in Rural Areas." *Rural Sociology* 59 (1994): 289–310.

Thompson, M.A. "Hurricane Katrina and Economic Loss: An Alternative Measure of Economic Activity." *Journal of Business Valuation and Economic Loss Analysis* 4 (2009): Article 5.

U.S. Bureau of Labor Statistics. "Business Employment Dynamic – First Quarter 2015." 2015. Accessed August 7, 2016. www.bls.gov/news.release/pdf/cewbd.pdf

U.S. Census Bureau. "Nonemployer Definitions." n.d. Accessed August 7, 2016. www.census.gov/epcd/nonemployer/view/define.html

U.S. Small Business Administration. "Understanding the HUBZone Program." n.d. www.sba.gov/content/understanding-hubzone-program

U.S. Small Business Administration. "U.S. Small Business Administration Office of Advocacy: Frequently Asked Questions." 2012. Accessed August 7, 2016. www.sba.gov/sites/default/files/FINAL%20FAQ%202012%20Sept%202012%20web.pdf

U.S. Small Business Administration. "Table of Small Business Size Standards Matched to North American Industry Classification System Codes." 2012. Accessed August 7, 2016. www.sba.gov/sites/default/files/files/Size_Standards_Table.pdf

Webb, G. R., Tierney, K.J. & Dahlhamer, J.M. "Businesses and Disasters: Empirical Patterns and Unanswered Questions." *Natural Hazards Review* 1 (2000): 83–90.

Webb, G. R., Tierney, K.J. & Dahlhamer, J.M. "Predicting Long-Term Business Recovery from Disaster: A Comparison of the Loma Prieta Earthquake and Hurricane Andrew." *Environmental Hazards* 4 (2002): 45–58.

Weick, K., Sutcliffe, K.M. & Obstfeld, D. "Organizing for High Reliability: Processes of Collective Mindfulness." In *Research in Organizational Behavior*, edited by R.S. Sutton & B.M. Staw, 81–123. Stanford: Jai Press, 1999.

World Economic Forum. "Global Risks 2015," 2015: http://reports.weforum.org/global-risks-2015/

Xiao, Y. & Peacock, W. "Do Hazard Mitigation and Preparedness Reduce Physical Damage to Businesses in Disasters: The Critical Role of Business Disaster Planning." *Natural Hazards Review* 15 (2014): 04014007-1-11.

Yoshida, K. & Deyle, R.E. "Determinants of Small Business Hazard Mitigation." *Natural Hazards Review* 6 (2005): 1–12.

Zhang, Y., Lindell, M. & Prater, C. "Vulnerability of Community Businesses to Environmental Disaster." *Disasters* 33 (2009): 38–57.

Zolin, R. & Kropp, R. "How Surviving Businesses Respond During and After a Major Disaster." *Journal of Business Continuity and Emergency Planning* 1 (2007): 183–199.

7 Managing Human Capital in Times of Crisis

The Role of Employees in Disaster Management

Stacey C. Mann and Jonathan W. Gaddy

Case Study: Calhoun County, Alabama

On April 27, 2011, Calhoun County, Alabama—like many other counties in the state—was impacted by a major tornado. The storm took nine lives in the county and generated millions of dollars in unbudgeted disaster-related work for the county in terms of debris removal, public facility restoration and reconstruction and disaster-recovery administration. The Calhoun County Commission was the most severely impacted of several local-government units in the county, which were facing the challenges of the disaster-recovery process and FEMA assistance in the aftermath of the tornado. County officials quickly realized the magnitude of the event and its potential for significant long-term impact lasting years into the future. Even while local emergency-management officials were coordinating massive volunteer emergency relief efforts, local-government administration in the county was working to secure an outside consultant—an experienced disaster-recovery administrator—to guide the county through the impending maze of FEMA guidelines and federal disaster procurement regulations. The disaster-recovery administrator was able to bring a level of experience and skill to the benefit of the county's citizens and treasury that it would not have otherwise possessed. Even so, all of the county's involved administrators and department managers quickly found themselves working double duty to meet the demands of the incident response and recovery while also striving to learn the relevant aspects of FEMA policy and procedure to ensure their work was carried out with the level of detail and care required by federal disaster regulations. Even with a devoted and skilled team combined with outside consultant help, the county still found itself making a number of temporary and part-time hires to perform disaster-related work, some of whom continued to work for more than a year after the disaster occurred.

The first priority for local government in a disaster is pivoting its resources and assets to meet the immediate life-safety needs of citizens—

this is the emergency-response phase. Coordinating search-and-rescue operations, providing temporary food and shelter to those displaced by the event and clearing a navigable path through debris-filled roadways are the immediate high-priority activities. During this immediate-response phase, the situation usually demands "all hands on deck," and local-government workers and the general public alike frequently find themselves working outside their day-to-day skill sets, such as law enforcement and public-works crews assisting with search-and-rescue duties typically handled by fire and EMS crews. Community volunteers also step in and can become de facto emergency responders, since the general public is always the first to arrive at the disaster-impact area.

Frequently after a disaster, local governments—typically with considerable assistance from non-profit organizations—set up volunteer-registration centers in affected areas to accommodate large numbers of spontaneous, unaffiliated volunteers (SUV's), or volunteers from the general public who are not affiliated with a recognized disaster-relief group but arrive at the affected areas to provide assistance. Organization and management of these SUV's are major logistical challenges but, when this is done effectively, the pool of SUV's can become a major force multiplier for overtaxed local governments, and especially their first responders. During the first few weeks after the April 27, 2011 storm, Calhoun County tracked and recorded over 4,000 volunteer man-days through its volunteer-registration centers. The convergence of these community resources with local-government efforts represents a post-disaster honeymoon period during which all efforts, public and private, tend to be focused on the common goal of providing humanitarian assistance to those affected by the event.

The tasks performed at volunteer-registration centers include determining work needs of first responders, identifying available volunteers to meet those needs and dispatching and managing teams of volunteers to perform work toward meeting local-government disaster-response objectives safely and with some measure of accountability. Another important task of volunteer registration-center personnel is ensuring that all applicable regulations and legal considerations are entertained, which includes educating volunteers on insurance policies that may or may not be available to them as they perform work, and providing safety briefings and just-in-time disaster job training prior to volunteers being dispatched to perform work. In Calhoun County, one of the most important goals of the volunteer registration-center personnel was to track the volunteers by name, the type of work performed by each, when and how long each worked and any special equipment (such as chainsaws or heavy equipment) used by the volunteer. For FEMA purposes, these gestures of goodwill on the part of the general public constitute donated resources and, if tracked adequately, can offset the local government's cost-share of total disaster-related expenditures (typically a 25 percent cost-share requirement that is sometimes shared between the state and the local-government unit). Donated labor and resources tracked through Calhoun County's process resulted in cost savings of over $600,000 toward the county's federal cost-share matching requirement, with the net result being that the county was not required to expend funds from its treasury to meet the match requirement.

A few weeks after the disaster impact, the process begins to look less like emergency response and starts to resemble what is known as the disaster-recovery phase. As the immediate needs of residents for rescue, housing and short-term assistance are met, volunteer and local-government goals begin to diverge as local nonprofits begin to articulate their roles and responsibilities for the long-term community-recovery process. Community volunteer groups that spontaneously emerge after a disaster may become permanent fixtures of the long-term recovery process as they work to continue delivering donated funds and other resources through the more organized long-term recovery, rebuilding and reconstruction funding mechanisms. Local-government goals shift toward recovery as well, as first responders return to regular work schedules and local officials consider how to coordinate community plans for rebuilding and how to oversee contracting for debris-removal and management activities.

In the long-term recovery phase, significant impacts on local-government human resources (HR) emerge. Local governments may be forced to outsource specific aspects of disaster recovery, such as administration of FEMA recovery grants and reconstruction of damaged infrastructure, due to insufficient staffing or organic knowledge. As local governments must continue to provide basic government services at a pre-disaster level throughout the response and recovery phases, these increased demands for administration of disaster-related work also can result in the hiring of additional part-time or full-time temporary personnel to augment existing local-government administrator positions.

Two aspects of local-government operations that often take significant hits in disaster recovery are financial management and grants administration. Financial managers must face the challenge of meeting tremendous unbudgeted disaster-related expenditures with often limited fund reserves. These disaster-related expenditures must be managed and executed in compliance with FEMA policy and guidelines, often requiring a significant amount of time investment in navigating the vast array of current policies and procedures related to procurement and financial management and in deciphering how to apply these policies to the local government's specific circumstances. This new knowledge also must be transmitted throughout the local-government organization to ensure that project managers in various departments are provided with clear, accurate guidance on how disaster-recovery work and contract projects should be accomplished in order to ensure maximum FEMA reimbursement for eligible expenditures. Understanding FEMA guidelines and federal-procurement regulations is typically outside the scope of many local-government managers. Occasionally, they are outside of emergency-management workers' breadth of knowledge as well. These personnel are tasked with executing the local-government recovery strategies, so a significant learning curve compounds the challenge of increased work volume for these officials.

A similar situation exists in grants management itself. Ensuring the local government takes advantage of available federal assistance through FEMA and other sources usually requires development of new record-keeping and retention systems in addition to the local government's existing filing and paperwork management processes. Managing this disaster-related paperwork is an example of the type of work often performed by disaster-related temporary hires. Cross-checking reimbursement documentation against FEMA policy and reconciling disaster-related claims with continually updated guidance from state and federal officials are ongoing processes throughout the disaster-recovery period, and can continue for two years or more before all claims are reconciled and reimbursed to the local government.

As governments at all levels witness policy change and begin to understand the value of their employees, the realization that human capital is an organization's most valuable asset continues to emerge as a major theme. As experienced in Calhoun County during the 2011 tornado, the human-capital component of disaster recovery is a significant factor for consideration by local-government officials at the onset of an event—if not before—as it will pose challenges to and present opportunities for the organization's effectiveness. Further, it will impact the ability to meet disaster-related demands for up to several years after the initiating disaster event. For the purposes of this chapter, "human capital" refers to the knowledge, skills, abilities, institutional memory and experience that employees bring to an organization, which can enhance resilience in a crisis.

Introduction

Local Governments, Human Capital and Disasters

Hurricane Katrina was a pivotal point in American history. The storm's effects on the Alabama, Mississippi and Louisiana coasts, as well as on the states that received thousands of evacuees, put unexpected demands on local governments. In those states devastated by the hurricane, local governments—although crippled—were still expected to function to meet the needs of citizens and businesses who were anxious to begin the rebuilding process. In the cities and states that received evacuees, local-government services were overwhelmed, as the demand for their services significantly increased. Many local-government employees were facing their own struggles at home, while others faced difficulties in what would seem a simple act—returning home. The result was that while local-government services were in demand, the supply of employees to assist in meeting those needs was limited.

In 2006, the Post-Katrina Emergency Management Reform Act included a provision that afforded the President the ability to approve the distribution of funds for employee salaries to local governments affected by disaster. Following a disaster, when employees are most needed, local governments' revenue intake decreases, which can affect employee salaries. When homes are destroyed by disasters, their value decreases, meaning that the property taxes gained by local governments also decrease. As Edwards and Afawubo report, the impact of lessened property taxes, along with other revenue sources such as sales and income taxes, has a significant effect on local governments, who are already on limited budgets with little extra (Edwards and Afawubo 2008, p. 87). Having to produce the extra dollars that are required to match federal and state funds for rebuilding becomes a difficult, if not impossible, task. Also, as the authors state, disasters can result in some local-government personnel incurring overtime expenses, further impacting local-government budgets.

Thus, local-government leaders must consider the implications of a disaster on the human capital within their own workforces. While some specialized positions, such as building inspectors and city engineers, require particular expertise, task re-orientation of other employees and utilization of community volunteers can often assist with bearing the burden of response and recovery. In fact, through the years, some governments have learned valuable lessons and have incorporated employment-related disaster policies into their emergency plans and employee handbooks.

Local governments also should consider the human capital available to them from the general public. As Dynes argues, in normal circumstances, being a citizen of a community simply means participating in the community, often on a minimal basis through activities such as voting or obeying the community's laws (Dynes 1968, p. 8). However, when disaster occurs, the citizen's role expands into one in which that person should assist other citizens and the community itself (Dynes 1968, p. 8). Because needs that were not present before a disaster may appear in its aftermath, local governments should be familiar with the human capital available to them outside of their own formal organizations.

The purpose of this chapter is to identify current practices regarding human-resources management practices (HRM) for times of crisis and effective methods of recognizing, developing and preserving human capital within local governments. The chapter will first provide a brief history of public human-resource management, followed by city and state policies regarding the use of employees during disasters. Then, using information from surveys with human-resource professionals and emergency managers in mid-size local governments, the authors will discuss best practices and lessons learned from disasters. Finally, policy recommendations regarding human-capital management and emergency planning, response and recovery will be offered.

Review of the Literature

A Brief History of Human-Resource Management in the Public Sector

The history of human-resource management in the U.S. is almost as unique as the story of the nation itself. The tides of change in public administration and policy have often been tied to reform in federal personnel policies, mandates and laws, which often trickle down to state and local governments (Nigro and Kellough 2008, p. 73). Hence, to best understand the human-capital approach to human-resource management that is now in existence, we must first understand the evolution from the personnel department to human-resource management. Although the personnel department was once viewed by bureaucrats and line managers as a group of "paper pushers" that held strictly to policies and procedures, the role of the evolved HR department has become one in which HR professionals assist with development, training and consulting. However, the road to finding their own chair at the bureaucratic table was long, sometimes challenging, and required that HR professionals evolve through professionalization and change within their own department.

In the first forty years of a young democracy, public servants were placed in office based on "fitness of character," which meant that the individual—recruited from the nation's elite societies—not only must have good character, but must also agree with the objectives and policies of the political party in office and be competent to fulfill the responsibilities of the position (Nigro and Kellough 2008, p. 18). However, as political parties began to organize and emerge and as Andrew Jackson took leadership of the country, a new sentiment began to develop—one that would assist top leadership with ensuring attainment of their goals. The motto "to the victors belong the spoils" accurately described this new system and the policy evolution in which "fitness of character" was no longer a prerequisite of being a public servant; rather, the victor was able to choose among those from any socio-economic class—those more representative of the electorate—to assist with the implementation and fulfillment of the political agenda (Goodman and Mann 2010, p. 184; Nigro and Kellough 2014, p. 18).

Although the implementation of the spoils system replaced only approximately 10 percent of public servants in federal government upon its initial implementation, the ebb and flow of government workers proved to be ineffective and inefficient in the minds of some. The Pendleton Act of 1883, which was enacted at the federal level and resulted in reforms in state and local governments, focused not on political patronage as the qualification for public-service positions, but rather on merit. The Act resulted in the creation of the United States Civil Service Commission, which also became responsible for administering competitive exams and managing position applications (Goodman and Mann 2010, p. 184). From here, state and local governments followed suit, emphasizing merit and procedures.

According to Lynn and Klingner (2010), efficiency in personnel management, as well as the allocation of federal jobs among the classes, became the focus of reform from 1883 to 1932; this was followed, from 1933 to 1964, by a model that focused not only on merit, but also on political patronage in the sense that the bureaucrat, or public servant, should be responsible to the political leadership. Political partisanship, especially in some positions, would assist with implementation of policies. However, after 1964, individual rights of employees as well as affirmative-action policies created a new atmosphere in which collective bargaining and social equity joined responsiveness and efficiency as major components to personnel management in the public sector (Lynn and Klingner 2010, p. 49).

Whereas public personnel management has endured reforms throughout the history of the U.S. government, policy change in recent years has emphasized federal workforce reduction, decentralization, performance and accountability, all of which transfer more responsibility to first-line managers. For example, the Civil Service Reform Act (CSRA) of 1978 created a system in which managers received not only "performance-based incentives" but also greater input in rewarding or disciplining employees based on their performance (Perry 2008, p. 202). The 1993 National Performance Review (NPR) also gave managers greater power and responsibility in "hiring, classifying, and assessing the performance of their employees" (Shiramizu and Singh 2006, p. 151). However, as witnessed in recent years with cuts due to an economic downturn, the NPR's focus on reducing the size of the federal workforce resulted in a smaller number of managers being responsible for more human-resource activities. Essentially, it was another modern case of having to do more with less.

This new approach to managing government, in which accountability, performance and decentralization became important managerial functions, emerged during the years leading up to the CSRA of 1978, and quickly became termed New Public Management (NPM). Like the NPR, the goal of NPM was to bring some private-sector practices to government (Box, Marshall, Reed, and Reed 2001, p. 611), in the hope to become more efficient and effective. Thompson and Miller (2003) argue:

> What the new public management calls for is the adoption of the organizational designs and practices that are transforming business: decentralized, flatter, perhaps smaller, organizations, structured around sets of generic value-creating processes and specific competencies, high-performance HRM practices, modern information technology, balanced responsibility budgeting and control systems, and loose alliances of networks.
>
> (p. 334)

As the federal government began to implement changes, so did state and local governments. In fact, Selden (2005) found that 25 percent of counties reformed some aspect of their HRM processes and that the largest refocusing of the HR department was toward strategic leadership, which the author described as "creating and implementing long-term goals, creating partnerships, and aligning human capital to meet the needs of the county government" (p. 63). Gabris, Drenell, and Kaatz (1998) found that local governments seemed to follow many of the federal changes, including reduction of the workforce, the move to employing generalists rather than specialists, emphasizing the team approach and focusing on accountability and performance, among things. Finally, French and Goodman (2012) reported that local government HRM had witnessed significant changes, especially toward New Public Management and a focus on "the human capital role," as now managers "have greater flexibility over their fiscal and human resources" (p. 76).

State governments also began to implement changes. Coggburn (2000) found that whereas states did not save money in personnel policy changes, the movement toward market-based approaches may have resulted in higher pay or managers motivating employees with economic incentives (p. 31). In addition, while greater efficiency and effectiveness were originally specific goals of HR reforms, Coggburn concluded that while the empirical data did not support that goal, the reforms did allow managers greater flexibility in decision-making. Nigro and Kellough (2008) also concluded that states have reformed personnel practices, but despite efforts to allow managers more control to meet organizational needs, the reforms have not necessarily created a trusting environment for employees, who once relied on the principles of merit for job security (pp. 55–6). Thus, the impact on trust may be to reduce employee motivation to assist managers with overall organizational goals. As stated by Shiramizu and Singh (2006, p. 152), "reform tactics should center on motivating people to improve themselves, which in turn will increase the productivity of the organization." However, as managers gain flexibility in meeting organizational goals, they also have an opportunity to understand and hone in on the human capital available to the organization. As employees begin to see that managers understand their talents, experiences and knowledge, their recognition of being valued also may lead to greater motivation and dedication to the goals and objectives of the organization as a whole.

Public human-resource management has endured much change through American history, and its policy evolution has had a significant impact on employees at all levels. Although downsizing and change in the merit system may have decreased trust among some public servants, the focus on human capital not only allows managers to acknowledge the skills, abilities and experiences of their employees, but also allows for greater development of those assets. During times of crisis, these qualities of employees—which are not always the talents that are included in their daily position's responsibilities—become important to response and recovery. Understanding the human capital available to the organization is critical, especially in mitigating and preparing for disasters. Thus, the lesson to be learned by public human resource-management history centers on the most important disaster-response asset—a government's employees and the knowledge, skills and abilities that constitute its human capital. As previous disasters have shown, effective management of human capital is the greatest response to disasters.

Personnel Reform as a Result of Crisis

In the years following the terrorist attacks of September 11, 2001, President George W. Bush created the Department of Homeland Security (DHS) to address potential crises. At its onset, the goal was to create a personnel policy for the agency in which managers had some flexibility, but which also protected employee rights (Ryan 2003, p. 103). One lesson learned from the disasters that occurred in the U.S. is that adaptability and flexibility to address threats and consequences are mandatory. Thus, managers need the same qualities with their own staff. The Bush Administration made personnel reform part of the agenda, and in the President's Management Agenda (PMA) in August 2001, one of the five proposals to address was the strategic management of human capital. The executive branch argued that the new DHS should have a personnel system that is "flexible, contemporary, and grounded in the public employment principles of merit and fitness" (Moynihan 2005, p. 182). Ryan (2003) wrote: "The administration's goal has been to design an agency as unencumbered as possible by traditional civil-service regulations arguing for best practices and a decidedly business school approach for the 'strategic management of human capital'" (p. 103).

According to Moynihan (2005), a major concern of the Government Accountability Office (GAO) was institutional memory, because the federal government was facing a significant increase in retirements (p. 175). Managing human capital in a way that would preserve institutional memory as a new department was formed offered a foundation from which to build, and offering managers flexible personnel policies in which they could be creative in maintaining that institutional memory would be pertinent. As implemented, the director of the Department of Homeland Security would have full discretion over personnel policies, which was argued to be necessary because "without such flexibility as a part of the broader security issue, the ability of the executive branch to effectively protect the public was compromised" (Moynihan, 2005, p. 179). Although critics argued that this policy would give the executive branch too much power, it allowed the director the opportunity to work with managers to identify personnel policies that could not only transfer institutional memory to the new department, but also allow managers to strategically utilize the human capital available to them, which cumbersome civil-service rules often prevented.

Public Law 107-296 of the 107th Congress, also known as the Homeland Security Act of 2002, served as the guide for the establishment of the DHS. In addition, the legislation served as a basic guideline for the new personnel policies. Because the focus was on "sustaining a culture that cultivates and develops a high performing workforce," the agency designated Agency Chief Human Capital Officers to oversee HR processes as well as the implementation of Strategic Human Capital Management (SHCM) (HSA 2002, p. 2289). Objectives of SHCM included:

- align human capital strategies of agencies with the missions, goals, and organizational objectives of those agencies;
- integrate those strategies into the budget and strategic plans of those agencies;
- ensure continuity of effective leadership through implementation of recruitment, development, and succession plans;
- sustain a culture that cultivates and develops a high performing workforce;
- develop and implement knowledge management strategy supported by appropriate investment in training and technology;
- hold managers and human resources officers accountable for efficient and effective human resources management in support of agency missions in accordance with merit system principles.

(HSA 2002, p. 2289)

At the agency level, FEMA's administration also noted the importance of strategic human-capital management, and specifically focused on recruitment, retention, training and compensation. According to a CRS Report for Congress on FEMA policy changes, the mistakes and "inadequacies in the number, deployment, and qualifications and training of FEMA employees" that were highlighted following Hurricane Katrina and Rita were the root cause of new personnel policies. Not only did FEMA stress the importance of training individuals with specialized knowledge, but it also allowed for payment of bonuses to retain "individuals in positions that are difficult to fill" (Bea et al. 2006, p. 24). At the end of each year, FEMA's administrator must submit an evaluation of the program "based on results-oriented performance measures" (Ibid, p. 25).

Although the Bush Administration's focus on decentralization, managerial flexibility and improved human-capital management may have sought to create better government, implementation of the reforms proved to be difficult. For example, collective bargaining rights became controversial when Democrats objected to the Republican stance that collective bargaining "would impede national security" and could potentially "obstruct effective government" (Naff, Riccucci, and Freyss 2013, p. 454). Naff et al. also point out that the U.S. Court of Appeals for the District of Columbia upheld a lower court decision that said the policy eradicated any "meaningful bargaining over fundamental working conditions" and that the DHS must "ensure collective bargaining over fundamental working conditions." In addition, the DHS created its own labor relations board, in which its employees' labor disputes would be heard, which the Court of Appeals ruled "illegally altered the role of the Federal Labor Relations Authority" (Rutzick 2005, p. 3). As a result of the changes, employee satisfaction within the department ranked near the bottom of all federal agencies, as did leadership and workplace performance, according to the Office of Personnel Management's 2006 federal workforce survey (Ballenstedt 2007, p. 5).

Although the reform movement within the DHS collectively was not largely popular and certainly faced its own share of problems, it is notable that the federal government understood that both planning for and responding to crises requires flexibility and adaptability. In addition, the realization that human capital and institutional memory are assets in mitigating, preparing for, responding to and recovering from disasters should be noted among state and local governments. In addition, these same characteristics that outside stakeholders, such as volunteer groups, citizens and the private sector, bring to the organization should be considered human capital upon which state and local governments can call as well.

While federal personnel such as FEMA disaster workers possess specialized knowledge about the disaster-recovery process, most local-government employees typically do not. This represents one part of the human-capital gap faced by local governments, which must quickly ramp up their organizational capacity to administer federal disaster-assistance funding and programs. Another part of the human-capital gap is in the larger community-oriented aspects of disaster response and recovery. As noted in the case study, volunteers from the community at large, such as local non-profit and community groups and even private businesses, can become force-multipliers by contributing to immediate disaster-relief needs in a coordinated manner. During the recovery process, these same groups can provide tremendous human capital in supporting long-term recovery and rebuilding programs, such as volunteer-based housing reconstruction initiatives (similar to Habitat for Humanity projects). In both phases, private-sector human capital may be directly engaged through targeted outsourcing, such as for disaster-recovery administration, debris management and removal services and long-term community-planning expertise.

Presently, many state and local governments understand the theory of human-capital management and, through mandates, legislation and policies, have implemented these standards for routine, day-to-day operations. However, the example of human capital in the public sector's day-to-day functioning also may be scalable to the realm of unanticipated disaster operations.

State and Local Government's Experience with Human-Capital Management and Disasters

In 1943, as a result of World War II and fear of an attack on the U.S. Pacific Coast, the state passed the War Powers Act as a measure to allow the governor to have additional emergency war powers. Government officials in California quickly realized that "human resources in California were inadequate to address the problems of mass attacks or natural disasters," so the California War Council was established to assist local war councils with recruiting volunteers who would train and serve during emergencies (OES 2001, p. 5). The Act also allowed workers' compensation to apply to volunteers injured while serving in the capacity of civil-defense worker. However, volunteers were not the only individuals who would be commissioned if an emergency occurred. In October 1950, the California State Legislature enacted the Levering Act, emergency legislation that not only required all California state employees to take an oath of allegiance to both California and the U.S., but also mandated that state employees would serve as "civil-defense workers" (Monroe 1952, p. 377). That term was eventually changed to "disaster service worker," which applies to all public employees who would are "subject to such disaster service activities as may be assigned to them by their supervisors or by law" (California Government Code 3100–3900).

The disaster service-worker legislation has proven to be a valuable commodity to California's local-government emergency managers, who have incorporated disaster-service workers into their emergency plans. When asked about the role of non-essential personnel, an emergency manager in California said, "All employees are considered essential as disaster-service workers during declared disaster." He added that employees have been assigned responsibilities that range from working in shelters to logistics and convergent volunteer management. An emergency-preparedness manager in another California city said, "We are a small city with a population of 100,000. All city employees are considered disaster-service workers and are cross-trained for critical roles during a disaster." Incorporating public employees into emergency plans not only better prepares the city for effective response if a disaster occurs, but also allows emergency managers and department heads the opportunity to discover other skills, knowledge and abilities that employees bring to the organization.

However, state governments also have long recognized the need for specific types of knowledge in local government, especially for disaster management, and, through the creation of legislation, mandates, policies and guidelines, local governments have been required to add skills or departments that require specialized knowledge and human capital. For instance, in 1933, the California Assembly passed the Riley Act, "which required all California local governments to have a building department and inspect new construction" (Butler 2012, pp. 43–4). The Act was passed as a result of changes in legislation for better construction of buildings following the 1933 Long Beach Earthquake.

Also in the early 1930s, as a result of the Dust Bowl, the federal government adopted legislation to improve farming practices that included soil conservation and erosion. States adopted the legislation and the federal government made funding available to assist local governments with the creation and operation of Soil Conservation Districts (SCDs), which would be directed by the Soil Conservation Service (SCS) (USDA n.d.; Hansen and Libecap 2004, p. 667). These districts are still in place in almost every county in the U.S. as a result of state-mandated legislation. Conservation specialists require unique knowledge of soil, including methods to prevent erosion, among other skills. The goal of SCDs is to help ensure that another Dust Bowl does not occur.

While these are only a few examples of ways in which legislation has impacted the investment of human capital, they also demonstrate that crises often pinpoint areas of knowledge or expertise that is needed. Although history provides local, state and federal governments many lessons, current practices provide not only glimpses of the past, but also insight to the future. The next section analyzes some of the current practices within local governments and how emergency managers across the U.S. are utilizing the human capital available within their organizations and communities.

Integrating Human-Capital Management and Local-Government Emergency Planning

As local governments reform and begin strategically managing human capital, HR departments also are transforming from administrative functions to strategic operations (Jacobsen, Sowa, and Lambright 2014, 295). In a 2014 nationwide survey of counties, Jacobsen, Sowa, and Lambright found that sixteen of forty counties had adopted a strategic-partner approach, and included the HR department in planning. Respondents reported that their governments realized the positive impact HR had on performance as well as the leadership characteristics of individuals within the department, which allowed others to understand the role that HR could play in the strategic direction of the organization.

Integrating human-capital management into the organization involves creating "an overall strategic plan as well as a human-capital plan that integrates the workforce requirements with the goals identified in the strategic plan" (Jacobsen, Sowa, and Lambright 2014, p. 291). Local governments also should have a guiding document in the form of an emergency operations plan (EOP) for addressing continuity of operations and response if disaster occurs. The National Response Framework, the federal government's guidelines for emergency response, requires a workforce that must fulfill the roles and requirements of the objectives of fifteen emergency support functions (ESFs). In creating a strategy for response, local governments must identify not only the resources available to the organization, but also those resources that are unavailable. Conducting a human-capital inventory may assist with identifying resources that are available, but may have been overlooked or unknown. From every crisis emerges a new set of obstacles, and while "essential" personnel provide consistent roles, the organization must understand the additional skills, knowledge and abilities that "non-essential" personnel can contribute to planning, response and recovery. With this knowledge, all individuals in the organization become "essential" in times of crisis. When surveyed, an emergency manager in Minnesota said: "We consider all of our employees essential during an emergency, kind of an all-hands-on-deck policy" (Mann 2013).

Transforming the organization to encompass human-capital management involves three operational requirements, all of which also are applicable to disaster management. According to Jacobsen, Sowa, and Lambright (2014), the three operational requirements include:

- identifying the strategic direction of the government unit;
- analyzing workforce requirements to achieve this strategic direction;
- developing action plans for the HR function that will help achieve the overall strategic goals of the department.

(p. 291)

To achieve each of these three capabilities, organizations must commit to planning. Emergency planning is an ongoing process. As the deputy fire chief in one Washington locale noted: "Exercise the plan! The best emergency plan has little value if it sits on a shelf." In other words, an emergency plan is a living, dynamic document that, once created, continues to be updated and practiced. To effectively prepare for crisis, an organization must consider all of its departments, the departments' needs during an event and the resources that departments can offer in response and recovery, all of which require planning, communication, cooperation, collaboration and decision-making, or careful strategy on the part of the organization as a whole. Organizational strategy, put simply, is pinpointing the organization's long-term goals, assessing the current state of the organization and its departments and, finally, identifying the actions required to reach that goal.

Organizational strategy also involves strategic agility, or the ability to provide "flexible, mindful responses to constantly changing environments" (Lewis, Andriopoulos, and Smith 2014, p. 60), a necessary component of effective response to disasters. To become strategically agile, organizations must have three capabilities. Lewis, Andriopoulos, and Smith point out that the first is strategic sensitivity, or the ability to be attentive to the environment and address problems with innovation and creativity. For example, during a disaster, affected organizations are faced with new and unusual tasks, such as debris removal, emergency public information and operating mass-care shelters. Strategic sensitivity is a critical skill for officials tasked with determining how the disaster-operating environment impacts the workforce, generates new work requirements and also provides new ways of meeting needs through volunteerism and disaster-recovery contracting.

The second required capability is leadership unity, which, for local governments, requires a collective commitment among managers, department heads and elected officials to collaborate and work toward the goals of the organization. Leadership unity is akin to the emergency-management concept of "unity of effort." Achieving unity of effort in a disaster requires rapid coordination of the various plans, objectives and strategies being generated by different disaster stakeholders and decision-makers at all levels and across all sectors, both public and private. To the greatest extent possible, emergency managers seek to maximize compatibility and shared goal-setting across all these organizations. A common, shared vision and set of disaster-response and -recovery objectives across all organizations is desirable because it allows officials to create dynamic partnerships and rapidly assess gaps and opportunities in their organization's planning and work efforts.

Finally, "resource fluidity requires change, switching, and novelty, but depends on consistency to take full advantage of resources" (Lewis, Andriopoulos, and Smith 2014, p. 61). Disaster-response and -recovery operations are dynamic and require ongoing re-evaluation and innovation.

For instance, workforces initially tasked to perform debris-removal functions may come from various agencies, such as the local public-works, fire and transportation departments. As roadways are cleared of debris and opened to emergency-response personnel, some departments may need to be transitioned to other duties, such as conducting building-damage inspections, while public-works crews may stay on to complete the debris-removal mission using heavy equipment. Later, public-works crews may disengage except for management and oversight of specialized debris-disposal crews contracted to perform final cleanup operations.

Analysis

HCM Best Practices in Local Government Emergency Planning

Because human-resource departments are responsible for many aspects of employee management, such as recruiting, compensation, benefits and training, their knowledge of and involvement in employee issues and knowledge, skills and abilities are vital to properly preparing local governments for potential crises. In addition, the knowledge HR professionals bring to the organization regarding human capital can prove beneficial when discussing not only important needs of employees during disasters, but also the resources they bring to the response.

In 2011, Mann conducted a nationwide survey of local-government HR professionals in U.S. cities with populations of 50,000–249,999 to investigate their involvement in emergency planning, as well as the inclusion of several human-capital components in local-government emergency plans. The study was conducted in January and February 2011 and was funded by the John C. Stennis Institute of Government at Mississippi State University. Approximately 605 potential respondents were contacted, and 213 completed the survey, resulting in a response rate of 35 percent.

In June and July 2013, a follow-up study, funded by Jacksonville State University's Office of Academic Affairs, was conducted in which emergency managers in the same cities were surveyed regarding their knowledge of HR involvement and human-capital issues important to emergency planning. Of the 605 potential respondents that were contacted, 213 completed the survey, resulting in a response rate of 35 percent.

Both human-resource professionals and emergency managers acknowledged the importance of human capital to their local government, especially in times of crisis. Although some local governments have taken deliberate steps toward implementing human-capital management, others may have knowingly, or even unknowingly, included elements of the approach.

The purpose of this chapter is to better understand current practices as well as concerns of the human-resource professionals and emergency managers who were surveyed in 2011 and 2013, and to identify ways in which emergency managers and HR professionals can assist chief administrative and executive officials in the process of managing human capital throughout the organization in times of crisis. The analysis will be developed using three objectives discussed by Tompkins (2002) and further developed by Jacobsen, Sowa, and Lambright (2014). These include: identifying the strategic direction of the government unit; analyzing workforce requirements to achieve strategic direction; and developing action plans for the HR function that will help achieve overall strategic goals.

HCM Requirement #1: Identifying the Strategic Direction of the Government Unit

Just as routine local-government operations require employees with expertise from a range of areas to fulfill the needs of the citizenry, so too does disaster response. From transportation, to debris management, to task re-orientation, the local-government unit must pivot in many dimensions in order to meet disaster-related needs while continuing to provide basic, day-to-day public services. Utilizing the emergency-planning team to more effectively identify strengths and weaknesses in this regard before a disaster, while also analyzing existing human-capital assets within the organization and the community at large, allows team members the opportunity to become acquainted and work together in the decision-making process prior to a potential event. This exercise could potentially result in the local government developing a clearer picture of what its own human capital may offer to the disaster-response and -recovery effort, and how best to outsource (or not) specific aspects of disaster-related work. More broadly, an organization can gain a more accurate impression of what the community at large may be expected to bring to bear to the benefit of the citizens through meeting their immediate needs while also offsetting the financial impact of a disaster on local-government coffers. Further, planning-team members are able to learn details about other departments on topics that range from culture, to language, to resources, broadening the understanding of the human capital available in ways not necessarily relevant to the day-to-day functioning of the government unit. The literature on organizational behavior and group decision-making has shown that "decisions made by groups of employees with diversified expertise will be higher in quality than those made by employees with more homogenous backgrounds" (Gruenfeld *et al.* 1996, p. 1).

As members work together, familiarity with one another increases. Research has shown that although conflict may still exist, groups who have worked or trained together are less likely to feel anxiety when expressing opposing opinions, are more likely to enjoy working together, are more likely to learn from others and tend to be more satisfied with final decisions (Gruenfeld et al. 1996, p. 1). In times of crisis, the trust built from previous interactions, the familiarity with other group members' expertise and the openness to express varying viewpoints can contribute to response. As demonstrated in Table 7.1, in the 2013 survey, emergency managers were given a list of twenty-three common local-government departments and asked to choose the departments that are included in the emergency-planning team. Answers ranged from one department to all twenty-three departments.

When asked important lessons about emergency-preparedness, several emergency managers stressed the importance of working with others throughout the organization:

> Be inclusive. Develop and tend to relationships—*Emergency-Management Coordinator, Arizona*

> Engage your Mayor and Council, City Manager, Chiefs and Directors—*Emergency Manager, Arizona*

> Working with others and getting buy-in from the beginning is important. Don't sit in your office and write a plan and then think everyone will agree—*Director of Emergency Management, Arkansas*

> Really sit down and think about how each of your identified hazards will affect your jurisdiction—everything from affected populations, to

Table 7.1 Local Government Departments and Potential Emergency Planning Members

Transportation	*Public Affairs*	*Communications*
Public Works	Human Services	Emergency Medical Services
Hazardous Materials	Public Safety	Finance/Administrative Services
City Council/County Commission	Economic Development	Board of Education
Firefighting	Housing	Engineering
Public Health	Search and Rescue	Agriculture/Natural Resources
Human Resources	Mayor's Office	Planning
Law Enforcement	Community Planning & Development	

debris management, to interruption of public services. Once you establish what can happen, you can then identify which person(s), organization(s) and stakeholder(s) need to be actively involved with your planning efforts, whether they have a seat at your planning table or are asked to review plan drafts—*Emergency-Management Coordinator, California*

Develop a strong team across departments and across agencies who share preparedness and planning missions as a priority—*Director of Public Safety and Community Relations, California*

Involve other city departments in the planning process—*Office of Emergency Services Coordinator, California*

Involve all stakeholders and [get] support from the top of the organization—*Disaster-Preparedness Coordinator, California*

If you don't want to be mired down when time is critical, ensure that all levels of the organization know they have a role in disaster management and are familiar with at least their piece of the mission. D-Day or Zero-Hour is not the time to begin learning—*Police Chief, California*

Start with a good team representing a variety of agencies—*Emergency Manager, Colorado*

To be successful in responding to emergencies, local governments must actively and consistently convene emergency-planning teams. As an emergency-management coordinator from Wisconsin said, it is important to "bring stakeholders together as much as possible, build relationships and exercise plans." As stated earlier, as members of the team become more familiar with one another, the more open they are to learning. Thus, frequency of planning meetings becomes important to the decision-making process. As indicated in Table 7.2, when asked how often emergency-planning meetings are held, emergency managers' answers ranged from "once a week" to "as needed." Many of the emergency managers reported meeting regularly, with 5 percent of respondents choosing once per week, 1 percent choosing every other week and 26 percent choosing once per month. Others reported meeting less regularly, with 5 percent choosing every other month, 16 percent choosing quarterly, 4 percent choosing bi-annually, 3 percent choosing annually and 25 percent choosing as needed. Meanwhile, 16 percent of emergency managers chose "other" and said that some other groups that join their local government emergency-planning team include parks and recreation, hospital associations,

Table 7.2 Frequency of Local Government Emergency Planning Meetings

Meeting times	# of respondents
Once per week	10
Every other week	3
Once per month	59
Every other month	11
Quarterly	34
Bi-annually	10
Annually	6
As needed	59
Never	0
Other	40

environmental health, utilities, gas companies and animal response, among others. Others offered valuable insight regarding the frequency of meetings with groups outside of the local-government planning team that also are considered stakeholders, who bring valuable resources to disaster response.

First, several respondents mentioned local emergency-planning committees (LEPCs), which are committees that were established by mandate of the United States Congress passage of the Emergency Planning and Community Right to Know Act, or Title III of the Superfund Amendments and Reauthorization Act of 1986 (SARA Title III). Upon enactment, the purpose of LEPCs was to bring together groups from throughout the community to create and update local-government emergency plans. According to Lindell (1994), the "variety of membership within the LEPC provides an opportunity for disparate parties to work together over an extended period of time in a noncrisis atmosphere that promotes the development of mutual trust" (p. 163). Although most of the time the specific mission of LEPCs was to concentrate on toxic chemicals, in some cities the groups expanded to examining other hazards. For example, one director of emergency management in Alabama said that LEPC meetings are held monthly, while ESF 8 meetings—or the National Response Framework's emergency support function 8, public health—are sometimes held monthly and at other times quarterly. Emergency managers in Georgia, Kansas, Oklahoma and Rhode Island report that their LEPC meetings are held quarterly.

While working on an emergency plan with members of other departments is pertinent, including outside groups strengthens the organizational strategy of local government because it offers an opportunity to further identify needs and resources within the community that may have been overlooked. Volunteer Organizations Active in Disasters (VOAD) and other volunteer groups were the most commonly mentioned, and the necessity of having someone directly appointed to manage the volunteers—

both with organizations and spontaneous—was a resounding theme. Several respondents also mentioned that the private sector and small businesses should not be overlooked. Hy and Waugh (1990) argue that the private sector should be a strong partner in emergency planning. For instance, because chemical companies help with detoxification and construction companies understand building codes, they bring expertise that might not otherwise be available. They point out that "Without help from the private sector in some cases, emergency management efforts are doomed to failure" (p. 14). Emergency managers nationwide agreed, and the emergency-management program manager from one city in Washington said that including local businesses and nonprofits will help with establishing a clear understanding of the needs and capabilities needed for effective economic recovery. The director of emergency management for one Wisconsin city added: "Involve all partners no matter the discipline, because the one you leave out will be the one that hurts you in the response and recovery phases."

Emergency managers also mentioned that they meet with regional and state partners. Emergency-management coordinators in two Colorado counties said that one to two local, regional and state meetings are held once or twice per week. A former emergency-management coordinator in Kansas said that the region is "heavily involved in the planning process" and planning meetings could occur up to four times per month, depending on the current planning topics. However, in terms of all partners, an emergency manager in New Mexico offered this advice: "Don't ask for too much from your partners—have realistic expectations of them." Emergency managers can expect (or at least hope), though, that as the responsible parties for their departments, department heads will "buy in" to the importance of planning and the positive effects that relationship-building and networking can offer. As leaders come together, inevitably problems faced by departments will be discussed, which may lead to innovative idea-sharing that benefits not only disaster response and recovery, but also daily operations.

Finally, some of these ideas also may be derived through tabletop exercises. In fact, practicing the plan and conducting regular training exercises were both important activities that were mentioned by many of the survey respondents. A Battalion Chief in Iowa said: "Working with other agencies/entities is vital [and] annual drills/tabletops can be very beneficial." In addition, several respondents said that learning lessons from previous incidents and incorporating lessons learned during exercises also are important. The emergency-management director for one city in Massachusetts said, "Learn from previous actions—both what went right and what went wrong or needs improvement."

Thus, to successfully identify the organizational strategy of a local government's emergency plan, the needs and capabilities of all departments, organizations and regional and state partners must be considered. Working with all stakeholders and holding frequent planning meetings and exercises allows for a thorough assessment of the current state of the organization and its departments, and the actions required to reach those goals. Further, understanding the capabilities of individuals, departments, organizations, agencies and others offers the core planning committee knowledge of capabilities that can be called upon when environments change. One of the characteristics of being strategically agile is adaptability to changing environments, so having individuals from various disciplines with diverse skills makes adaptation in many circumstances easier and more viable. However, failing to bring these diverse partners together frequently can be a costly mistake. As a Michigan director of public safety said: "Plans are worthless, but planning is invaluable."

HCM Requirement #2: Analyzing Workforce Requirements to Achieve Strategic Direction

The local-government workforce is the most valuable asset to local government as well as to the community, because the employees will help to meet the final strategic goal of effective response. Thus, having a clear understanding of the workforce capabilities will lead local governments toward achieving their strategic directions. Understanding the capabilities of the workforce begins at the department level. Department heads are responsible for understanding the needs and capabilities of their employees, and in terms of disaster management, one way to identify these characteristics is through an emergency plan for each department. When emergency managers were asked if individual departments within their organizations had emergency plans, 55 percent replied that most departments have their own plans, 30 percent answered some, 12 percent answered a few and 3 percent answered none.

Of course, one of the most important components of emergency planning at any level is identification of the human capital both available and unavailable to the department and the organization. As leaders, department heads must understand some components of human-capital management that are important to the overall organizational planning process. First and foremost, an organizational structure needs to be clearly defined. Will the county administrator serve as incident manager? Will the police chief? Will the emergency manager? How will roles and responsibilities be distributed across each of these positions? How will this distribution evolve over time as the disaster progresses from response to recovery? Creating an organizational structure prior to an event produces three positive results. First, those who serve as leaders of the organization, some of whom may not ordinarily have leadership positions, will understand their role. Second, creating an organizational structure will assist with identifying the human capital needed, such as persons with knowledge of commanding the incident, experience with working with the media and familiarity with volunteer management. Finally, personnel throughout the organization will know who to report to, as well as the appropriate individuals that will handle problems as they arise. Among the 213 emergency managers who completed the survey, 90 percent reported having an organizational structure in place for times of crisis. Many local-government leaders often have similar roles in times of crisis, which allows for the use of skills and abilities that they use typically. For instance, one director of emergency management in Rhode Island said: "Most employees with a supervisory, leadership role are assigned to the EOC."

Leaders of the organization are commonly considered essential personnel, or individuals in positions that are pertinent to disaster response. These individuals possess particular skills or knowledge for the position they fill not only during times of normalcy, but also during a crisis. Essential personnel should be identified prior to the crisis so that they can report directly to their assigned locations and so that they are prepared to take on the task. "Responses need to be automatic given the high levels of stress during the crisis" (Cavanaugh 2006, pp. 5–6). Fortunately, 83 percent of the respondents indicated that their locales currently have identified pre- and post-disaster roles of essential personnel beyond their regular duties. In addition, 78 percent said the essential resources needed for response workers have been identified, 68 percent have identified primary or secondary meeting locations for employees in the aftermath of disaster and 90 percent have created employee contact lists.

Some employees may be considered non-essential, but still have much to offer the organization during crises. A director of emergency management in Georgia said, "We no longer use the term 'non-essential.' This implies 'not needed'; other terms would be more fitting, such as 'team leaders' and 'support personnel'. Support personnel handle all other functions not handled by the team leaders." Many emergency managers reported that non-essential personnel work in a variety of roles, including working in the EOC, assessing damage, managing volunteers, serving at points of distribution and supporting other personnel. One California deputy fire marshal said, "We train every city employee on awareness of emergency operations and inform non-essential personnel of their potential roles. We run a very lean staff and will need every person to assist at some level." Similarly, an emergency-services coordinator in California said,

> All of the employees are considered Disaster Service Workers in the event of a disaster; therefore, they are required to report to work. Cross-training happens because it is likely that when a disaster happens, employees will not be doing their 'normal' job and need to be flexible.

Although 64 percent of the survey respondents agreed or strongly agreed with the statement "For times of crisis, my local government has implemented cross-training for employees in preparation for emergencies," 35 percent neither agreed nor disagreed, 15 percent disagreed or strongly disagreed and 21 percent neither agreed or disagreed or did not know.

In the leadership role within their own departments and as department representatives at emergency-planning meetings, department heads can assist with several aspects of planning regarding non-essential personnel. First, by understanding the needs of the department, the department head can identify roles or responsibilities that may need additional assistance during crises. Further, the department head also can identify the knowledge, skills and abilities that may be applicable to other departments in disaster response and recovery. For instance, an employee on the library staff may have previously held a position as a school-bus driver, and would be able to assist with evacuation of special-needs individuals. In creating their own emergency plans and to further contribute to the emergency-planning strategy of the organization, department heads may consider conducting a skills inventory of current employees. Conducting a skills-inventory form may reveal knowledge, skills and abilities of employees that may not have been identified otherwise. The inventory also may quickly reveal that an individual who once was considered support personnel may be re-classified to essential based on previous experience. Fortunately, 63 percent of survey respondents agreed or strongly agreed with the statement "The emergency plan for my local government includes identification of pre- and post-disaster roles of non-essential personnel (beyond regular duties)."

However, to effectively use support personnel, both department heads and local-government officials must identify a couple of important policies. First, support personnel must know when and where to report, both of which Goodman and Mann (2008) identified as major issues for some local governments in Mississippi following Hurricane Katrina. Among survey respondents, 53 percent said their local government emergency plans include a timeline for non-essential or support personnel to report to work. Thus, sharing primary and secondary meeting locations with support personnel is essential, as well as clarifying expectations for reporting to those locations.

Finally, as discussed previously, working with groups and organizations outside of local government is vital to emergency response. Whether individuals are associated with a particular volunteer agency or are spontaneous volunteers who arrive at the disaster site to offer assistance, volunteers will come. Thus, local governments must ensure that the management of volunteers is included in the emergency plan. Because emergency managers have learned lessons either from previous incidents or from other jurisdictions, volunteer management has become a priority to many local governments. In fact, not only did 59 percent of emergency managers report that their local governments had identified methods for volunteer management, but 76 percent reported that their local governments established procedures for reporting volunteer injuries in times of crisis.

Human-capital management is vital to emergency planning, and as survey respondents have indicated, HCM has become a top priority among local-government emergency-planning teams. Understanding the workforce required to effectively respond to disasters is only one aspect of the emergency-planning process, but clearly, it is an aspect that cannot be overlooked. As previously discussed, employees are the most vital asset of the organization and the community, so including the department that is most knowledgeable about employee-management issues—the human-resources department—can be a vital asset to the emergency-planning team. The next section will discuss ways in which HR can assist with achieving goals of both the department and the organization.

HCM Requirement #3: Developing Action Plans for the HR Function that will Help Achieve Overall Strategic Goals

Because the third requirement focuses on HR, it becomes vital for HR professionals to not only understand the emergency-planning process, but also witness the process as it occurs. Thus, including a representative of the HR department on the planning team achieves these goals. Of the 213 HR professionals who were asked if a member of the HR department regularly participates in emergency-planning meetings, 68 percent answered yes; similarly, 67 percent of emergency managers said that an HR representative is included as a member of the response team.

Fortunately, many local governments understand the value of human-resource professionals in disaster management and have included HR representatives in assisting with strengthening response. Volunteer management has become an important task, and several emergency managers said that the HR department could bring valuable skills and knowledge to that function. For instance, the management of volunteers also could also entail reviewing skillsets and placing volunteers in areas that are appropriate and in need of them.

Likewise, several emergency managers said that HR professionals understand the skillsets of current employees, and an emergency-services planner from Arizona said HR professionals could assist with "how employees will be utilized in emergency response and recovery if they do not have a specifically designated role." Because HR professionals understand the needs and demands of the organization and are able to assess the current workforce, understanding the value of diverse skillsets and their applicability during a disaster is an important HR function that could assist with task re-orientation and cross-training. An HR director in California added:

> HR professionals should not only have updated lists of available resources, but also have an idea of what types of positions they will be filling, and how to schedule so that each position is filled at least three days in advance for each 12-hour shift. It's not enough to know all the KSAs [knowledge, skills and abilities] of your resources. You should have awareness of how those KSAs would be used in a disaster so you can anticipate the need. The tasks assigned in responding to an emergency are sometimes very different than our day-to-day jobs.

Because responsibilities during emergencies are often dissimilar to daily positions, several emergency managers said that HR could provide applicable training. Several emergency managers said that HR could assist with training current and future employees with the NIMS (or National Incident Management System) and ICS (Incident Command System), both used during disaster response. The same HR director from California added: "All HR professionals should take all necessary ICS training to understand the reporting structure, and to educate themselves on all types of disasters." HR professionals also can assist with training or scheduling training, such as FEMA training for volunteer reception.

However, before some types of training can occur, HR professionals can assist with assessment of the current workforce. Forecasting the needs of the organization, in a manner also informed by a forecast of the needs of the community at large, and providing training to fulfill those needs are important HR functions that are vital to emergency response. Because these are important skillsets of HR professionals, their inclusion in emergency planning is pertinent. In fact, 32 percent of HR professionals said that they conduct regular assessments of employee knowledge, skills and abilities useful for disaster response and recovery, and 66 percent said they continually assess internal workforce availability. In addition, 43 percent of HR professionals indicated that they regularly forecast the internal and external supply of employees. Forecasting, especially for the public sector, becomes an important issue when considering the effects of a major disaster on the workforce within the community. In fact, the supply of employees may significantly decrease following a major disaster for a variety of reasons, including lack of housing, better pay in the private sector or relocation (Goodman and Mann 2008, pp. 10–13).

Further, like the community surrounding them, public employees often face the same hardships as their neighbors, with their homes damaged or destroyed and their families in need. The emotional toll that employees endure as they assist with recovery at work, then at home, can result in significant stress. HR directors were asked if counseling for employees is mandated during states of emergency, and 14 percent said counseling is mandated in their local government. Unfortunately, intense stress and other psychological factors can lead to increased absenteeism in the workplace. Following Hurricane Katrina, representatives of Mississippi local governments said their emergency plans did not include policies regarding employee furloughs and discipline to address some of the problems that surfaced (Goodman and Mann 2008, p. 6). However, Mississippi is not alone concerning these policies when compared to the nation. In fact, 56 percent of HR directors reported that they do not have policies regarding employee retention after a disaster, 66 percent said they do not have policies for disciplining and terminating employees during states of emergency and 55 percent said they do not have policies that address employee furloughs during states of emergency. With this in mind, local governments also should consider hiring policies—such as for temporary workers—during times of crisis. An HR specialist in Texas agreed and said local government must consider including "procedures for accounting for absence due to the disaster—in other words, if a full-time employee cannot come to work due to the event, must their unworked time be charged to Annual Leave? Sick leave? Unpaid leave?"

Therefore, local governments must consider how the skillsets that absent employees would bring to the effort would be replaced. Although only 36 percent of HR professionals reported having policies that address hiring practices during states of emergency, 79 percent reported that their local governments have policies that allow for the immediate hiring of temporary workers.

While understanding the skillsets needed and available in times of emergency is important, local-government compensation policies that apply to their current employees also must be considered. As an HR director from Colorado commented, "Payroll fits into HR. How we would pay people if our systems went down is a big issue." Thus, HR directors were asked if they had identified a partner outside of the region that could help with IT issues, such as the necessary software for running payroll. Unfortunately, only 31 percent of HR professionals said they had identified another organization to help with these issues. However, 65 percent reported having analyzed potential compensation issues that may arise post-disaster such as overtime and disaster pay.

With these issues and policies in mind, clearly local-government HR departments are valuable commodities to both emergency planning and response. Although many local governments have not considered some of these important policies, others reported that they had current efforts underway or that as a result of the survey they would begin considering the importance of these issues during times of crisis.

Findings

The U.S. government has endured constant reform since its inception, particularly in the area of human-resource management. However, the goal of each and every reform has always been the same—to hire the best employees to provide the best services for the citizens, especially in the worst of times. Responding to and recovering from disaster have been as much a part of American history as the constant reformation of public policy, and often that policy reform has been a consequence of disaster.

Whether at the federal, state or local government levels, disasters bring new perspectives on ways to better prepare, respond and recover, and also how to better identify and manage capabilities and resources. While managing human capital for disasters is not their primary responsibility, emergency managers and human-resource professionals bring valuable knowledge to the process. Because they understand the potential impact disasters bring, emergency managers are able to provide knowledgeable insight on possible personnel needs. Human-resource professionals are extraordinary assets because they understand the human capital available to the organization from within, as well as potential human capital available within the community. When combining their expertise and experience, better preparedness measures may allow for smoother, more effective response to crises.

Among those who responded to the 2011 and 2013 surveys, both emergency managers and HR professionals saw the benefits of collaboration for further preparedness in local government. By working together, these partners can assist administrative officials and individual departments with examining human capital, which can result in significantly better response and recovery actions during and after disasters. Successful adaptation to small crises and unexpected consequences that occur during disasters is more likely to occur if all local-government departments have worked together and have a clear understanding of the capabilities and resources available to them. The results of the combined efforts lead to greater communication, effective collaboration and continued learning, which emerged from the surveys as key themes.

Communication issues often plague disaster response, so opening lines between departments and ensuring clear transmission can not only assist with adapting to situations that surface, but also potentially save lives. For example, as representatives from each department convene at emergency-planning meetings, communication among the entire organization will be initiated, and as those representatives begin to analyze the human capital within the individual unit, communication in individual departments also will be initiated. As individuals within departments begin to discuss the knowledge, skills and abilities they can bring to the organization, the strategic direction of the unit can be better defined, which will result in a clearer vision of the department in the overall organizational strategy.

Further, as emergency managers, HR professionals and department heads begin to examine the resources and capabilities within their organization for disaster response and recovery, they also will begin to identify the gaps. Thus, developing relationships with outside partners will not only bring a multitude of resources, but also open communication with stakeholders that could be affected by a disaster. In turn, as relationships strengthen both inside and outside the organization, communication failures that are often witnessed when disaster occurs are less likely, and more importantly, the amount and variety of human capital available to the organization increases. As a division chief in a Florida fire department said, "Communication and coordination of resources is key."

Collaboration and the creation and maintenance of relationships also were key themes that survey respondents consistently reported as important, and fortunately, 65 percent of emergency managers reported that authorities in their local governments supported such efforts. In managing human capital, then, collaboration begins with emergency managers and human-resource professionals, who can hold one another accountable in their commitment to the long-term planning process. Upholding the commitment to collaborate also provides a good example for department heads, who will contribute to human-capital management by working with their staff members to identify knowledge, skills and abilities outside of their current job responsibilities. Meeting regularly to discuss the progress within each department and assisting others in the process creates a learning environment in which trust builds and collaboration can grow. When asked which lesson learned would be the most important to share with other emergency managers, the director of public safety in one California city said, "Develop a strong team across departments and across agencies who share preparedness and planning as a priority." As the department that interacts with all organization employees and understands human-capital management, the HR department, as one Washington fire chief put it, "could be the catalyst to get the ball rolling between departments." Thus, by collaborating with the emergency manager to understand the needs of the organization, followed by working with department heads to identify the skills available, HR professionals offer a valuable lesson on how collaboration with all departments can benefit the organization as a whole.

Finally, education and training were important components of emergency planning that many emergency managers and HR professionals said should not be overlooked. An HR analyst in California said that during a disaster, personnel from every department serve on the emergency-response team, so regular response training is necessary. A deputy director of HR in another California city said that the HR department is responsible for providing emergency-preparedness training such as CPR, CERT and state-recognized communication courses, and that through updated training throughout the year, employees understand their roles and the tools available to them. Emergency managers throughout the nation agreed, and many said that the human-resource department could be valuable in identifying the skills available from employees organization-wide and then providing specific training for those skills in disaster.

Because HR professionals focus on employee-management issues and emergency managers understand disaster management, bringing the two together allows for matching an employee with important roles before, during and after a disaster. Communication, collaboration and education are important aspects of an effective organization, both during times of normalcy and in times of disaster, and allowing emergency-management and human-resource management to pursue these policies can result in positive consequences when disaster strikes. However, as with many local-government departments, emergency managers and human-resource professionals must focus on those projects that they or their supervisors consider to be dire or the greatest priority. Often, in times of normalcy, emergency planning does not qualify as either.

When asked about obstacles that they felt hindered effectiveness in their positions, 82 percent of emergency managers said lack of staff, 74 percent said lack of funds and 71 percent said lack of time. Emergency managers have a variety of duties and responsibilities, which might include creating and updating plans for the municipality, working with other jurisdictions for regional planning and attending and presenting information to the public through outreach programs. In addition, many of the respondents who serve as emergency managers in their respective jurisdictions also have roles in other departments, such as police and fire departments. In fact, nearly 40 percent of emergency managers surveyed have responsibilities in addition to their emergency-management duties. However, as an emergency manager in California said, it is important to "keep up the effort despite the 'busy-ness' of day-to-day work requirements throughout the organization." Sharing some of the tasks with other departments who can examine how they would be beneficial in disaster response also can lessen the load.

Human-resource management personnel also face similar obstacles; however, several HR professionals stressed the importance of emergency planning, and agreed that human-capital management should be a priority. For example, an HR director in Illinois said:

> We've been lucky in that our fire and police are very well-structured during emergencies. Not so on the HR side. I've only been in this office a few months and my predecessor did not actively engage in emergency planning. I discovered during the blizzard that we were woefully unprepared. HR professionals need to take this planning as seriously as our police and fire professionals do!

Although funding for training is usually the first item cut when budgets are tight, an HR manager in Utah reported that grants are available to hire specialists who can provide the proper training, and some foundational emergency-management training such as ICS and NIMS courses is freely available on FEMA's website. To overcome the time obstacle, several HR departments reported including some preparedness information in their new employee-orientation programs. Finally, as mentioned previously, collaborating with other department heads and emergency management, which HR professionals said is important to the process of planning, will assist with sharing the workload.

Another potential problem that could surface when devoting significant resources and time to managing human capital for times of disaster is lack of support by local-government politicians. Although only 25 percent of emergency managers felt that lack of local support hindered their position, several emergency managers and HR professionals mentioned the importance of local politicians' support. In particular, one Florida HR manager said,

> Get the buy-in of the administration and appointed officials. Get the Emergency Management Coordinator in front of the leadership of the organization on a regular basis to emphasize the importance of the maintaining the plan even when an event has not occurred in a number of years.

Political support also could transition into additional resources, especially if they have full understanding of the impact of being unprepared. Discussing the potential impacts of the major hazards and worst-case scenarios faced by jurisdictions can offer some insight about the consequences of not devoting time, effort and money to emergency planning. An emergency-management coordinator in California gave this advice:

> Really sit down and think about how each of your identified hazards will affect your jurisdiction—everything from affected populations to debris management to interruption of public services. Once you establish what can happen, you can then identify which person(s), organization(s), and stakeholder(s) need to be actively involved with your planning efforts, whether they have a seat at your planning table or are asked to review plan drafts.

Although many politicians may not commit to attending planning meetings, reviewing drafts and discussing their roles are methods of ensuring their involvement. In addition, ensuring that politicians know they have a role during and after disasters and specifically defining those roles can help them understand the importance of planning. As a Maryland emergency-management coordinator said, "Get the top government officials or their representatives actively involved. This will encourage participation at all levels."

Conclusion

Managing human capital is a responsibility that includes the work of the entire organization. Emergency managers are not endowed with responsibility for human-capital decisions, despite being responsible for allocating and reconfiguring human capital during and after an emergency/disaster. Ultimately, the responsibility for human capital resides with chief administrative and executive officials, who may be constrained, both cognitively and politically, from supporting or driving the kinds of human-capital reallocations needed for the local-government unit to best meet the needs of a disaster.

Solving—or, rather, preventing—the problems of human-capital management in a crisis is not simply a matter of inviting the emergency manager "to the table" with human-resources officers and administrators/executives. Rather, department heads, administrators, executives and human-resource managers should understand the factors affecting the re-configurability of the local-government unit's human capital to support disaster needs. Further, they should participate in and support the necessary pre-incident planning and trans-incident personnel moves and job assignment changes necessary to support a "whole of government" approach to responding to the incident. This involves both educating and empowering these non-EM officials and negotiating decision points and acceptable limits of human capital-management strategies pre-incident, and, subsequently, the enforcement of these plans and agreements trans-incident by the chief executive of the local-government unit.

To accomplish the requirements of successful human capital management, three overarching recommendations have been derived from the 2011 and 2013 surveys. First, local governments should ensure that every department is represented at emergency-planning meetings. From parks and recreation to transportation, every department has not only a stake in the process, but also potentially untapped human capital vital to planning, response and recovery. Through collaboration between emergency management, human-resource management and other departments, local-government officials can ensure that they are providing the best services at the time they are most needed.

Second, an environment of preparedness should be encouraged throughout the organization. Department heads should encourage employee participation as well as ensuring employees understand that all of their skills, including those that may not be used regularly in their daily positions, are potentially assets to disaster response and recovery. In addition, administrators and officials should stress that during times of crisis, all personnel are essential to the organization. Although the smaller municipalities represented in this survey commented on the importance of including every individual in the organization, larger governments have greater numbers of residents to serve, requiring more employees to assist.

Finally, building relationships with individuals, groups and organizations outside of local government can bring valuable skills, knowledge and abilities to the planning and response processes. Disasters affect the entire community, and having skillsets from various backgrounds and disciplines available may prove to be the most significant resource when responders face unforeseen consequences. Thus, individuals within the organization must be trained to identify and task the human capital that becomes available from spontaneous volunteers.

Disasters stress governments in often unanticipated ways and stress organizations' human capital through shifting job responsibilities and overtime required to respond to the incident. Even when dealing with the extraordinary circumstances brought on by a disaster, governments must continue to meet the routine needs of their citizens. As a result, some governments have incorporated human-capital issues into pre-disaster planning efforts. Governments can also benefit by recognizing the important role volunteers play in increasing the human-capital pool available for disaster-response and -recovery efforts.

Human-resource management in the U.S. has evolved from early systems of patronage to a professional, modern human-capital approach. The need for a professional workforce increased along with the complexity of government; as a result, an environment of meritocracy was cultivated through the implementation of civil-service systems across all levels of government. As government human-resource management continued to evolve, efforts to safeguard employee rights while preserving the executive discretion of elected officials resulted in the emergence of collective bargaining in the public sector, along with initiatives designed to limit discrimination in personnel-management decision-making.

Government personnel managers took lessons from the private sector as they worked to improve accountability and enhance performance in their organizations. One of these lessons was to incorporate deliberate human-resource planning into larger bureaucratic planning and development efforts to ensure government needs were more closely aligned with human resource-management efforts. Adopting a view of personnel in terms of "human capital" resulted in efforts to further personnel development, recruitment and assignment that acknowledged the critical role played by personnel in allowing government to meet the needs of its citizens.

In the wake of the September 11, 2001 attacks, the creation of the Department of Homeland Security brought significant challenges in human-capital management for the federal government. The merger of many separate agencies into a new, integrated organization afforded an opportunity for policy-makers to implement best practices from the private sector in strategic human-capital management. Partly as a result of the exigent circumstances of its creation, the Department of Homeland Security was afforded broad discretion in its ability to manage human capital within its own domain, in contrast to the restrictions imposed by the merit systems under which other federal agencies operated. The agency designated human-capital officers to oversee implementation of key strategic human capital-management efforts. In spite of these efforts, a few years later, FEMA came under fire in the aftermath of Hurricane Katrina and Rita for alleged poor performance. Many of the agency's failings were attributed to poor human-resource management, especially regarding employee training, recruitment and retention. As the Department of Homeland Security continued to evolve, court decisions criticized the department's reliance on its own internal labor relations board. Continuing struggles related to human-capital management and associated reform efforts within the department contributed to a poor working environment, with the department ranking near the bottom on metrics of satisfaction, leadership and workplace performance relative to other federal agencies. Despite these challenges, it should be noted that the federal government expressed understanding of the important role for flexibility and adaptability in a department focused on disaster and crisis-management activities.

In contrast to the employees of the federal Department of Homeland Security, most local-government employees typically lack specialized knowledge of disaster-response and -recovery processes. Local governments, partly due to their relatively small size, may rely extensively on community volunteers, non-governmental organizations and private-sector partners to augment their workforces in the aftermath of a disaster. Most of the rules governing personnel management in local governments are focused on day-to-day operations, but some state and local governments have begun to address the unique needs brought on by disasters in their policies, procedures and regulations.

During and after World War II, in the face of a possible foreign invasion, the State of California enacted laws that required public employees to be available as civil-defense workers in the event of an emergency. These laws have evolved over time and now constitute the state's disaster-service worker program. Local-government emergency managers in California incorporate disaster-service workers in emergency-planning efforts as a means to pre-plan human-capital employment to meet disaster needs. Typically, planning efforts related to the disaster-service worker program involve cross-training public employees to perform specific, disaster-related duties. Some legislation has also focused on the specialized skills required for disaster operations. For instance, California requires all local governments to employ building inspectors to ensure construction activities are compliant with the state's earthquake-damage efforts. Similarly, the federal government enacted measures such as the creation of Soil Conservation Districts to prevent a recurrence of the 1930's Dust Bowl disaster.

Deliberate human resource-management efforts have a positive impact on employee performance and employee development. Aligning human-resource efforts with overall strategic planning at the organizational level is a central theme of strategic human-capital management. Local governments should also align human-capital strategy with emergency and disaster strategy, such as that set forth in local-government emergency operations and disaster-recovery plans.

A critical capability of local governments in emergencies and disasters is the efficient re-tasking of government personnel to meet disaster needs. One tool that may serve to support this capability is the performance of a human-capital inventory that identifies potentially untapped resources that may be provided by a particular personnel pool, such as an employee's knowledge of special disaster-related skills that are not part of the employee's day-to-day job responsibilities. In order to align human-capital management with emergency-management activities, planners should consider how existing government responsibilities, as exercised through components and departments, will be stressed, and identify personnel pools likely to be available for re-tasking to other disaster-related duties. Ensuring the effective implementation of these plans requires achieving agility and unity of effort on the part of local-government leaders, who must actively work to overcome barriers to the implementation of disaster-related human-capital management tasks.

A 2013 study revealed that both human-resource professionals and emergency managers in local governments understand the importance of human capital to their local government's activities. In order to support these activities in a disaster, emergency-planning teams should include human-resource officers who can identify barriers and opportunities to effective employment of human capital in a disaster and assist in planning to overcome those barriers. An important activity of planning teams is relationship-building, which enhances an organization's ability to make decisions in a crisis. Emergency-planning teams should also convene regularly in order to ensure special disaster-related issues remain salient among team participants as they perform their routine, day-to-day job functions. A variety of examples exist in terms of local planning teams, including Local Emergency Planning Committees (LEPCs), created under federal environmental protection laws, as well as Voluntary Organizations Active in Disaster committees, formed to align and deconflict interests among non-governmental organizations and volunteer groups in times of disaster. Emergency managers' and human-capital managers' participation in these type of planning teams is critical, as it serves to build foundations for how government, community volunteers and the private and non-profit sectors may coordinate efforts in times of crisis. Likewise, it is important to stress the critical role of dialogue among officials from all levels of government—both within local-government units and with their state and federal partners—in comprehensive disaster-planning efforts.

The local government's workforce is its most valuable asset. In order to use that asset, local-government leaders must have a clear understanding of the workforce's capabilities and a method for coordinating the workforce to implement those capabilities. Responsibilities for workforce management are distributed among supervisors, department heads, administrators and executives. However, study results show that not all government departments and components have dedicated emergency plans. A critical concern during crisis is continuity of leadership. Addressing who will be in charge of a disaster before the event occurs serves to align workforce efforts quickly and efficiently with the needs of local government to quickly adapt and respond to crises. Study participants indicated that most have a pre-defined organizational structure for use in times of crisis. A majority of study respondents also indicated that associated planning measures include the designation of defined disaster-related roles for local-government employees, understanding the essential resources needed for emergency-response workers, identification of alternate work-reporting locations for employees and the creation of employee contact lists. Some local-government workers may not have pre-defined disaster roles and responsibilities. Survey respondents also indicated that pre-disaster cross-training of employees is conducted for emergency-preparedness purposes.

It is important for leaders to understand their own special roles and responsibilities in times of crisis. Department heads should be familiar with their own duties, as well as any special capabilities of their personnel that may go above and beyond duties performed by those employees on a routine basis. In order to capitalize on this knowledge, leaders should develop policies on the utilization of non-essential and support workers in a disaster. Study participants indicated that just over half of the represented local governments included procedures on work reporting for non-essential and support personnel. Likewise, local-government leaders should plan ahead for the important role to be played by community volunteers in disaster-response and -recovery efforts. Study results indicate that a majority of local governments surveyed have developed plans and procedures for volunteer management.

To address the concerns outlined above, it is important for local-government human-resource officials to participate in emergency-planning efforts. Among human-resource officials surveyed, two-thirds indicated that they participate in emergency-planning meetings and that human-resources personnel participate in their organization's emergency-response team. Emergency managers have indicated that one important role to be played by human-resource specialists is in the management and effective utilization of volunteers. Similarly, emergency managers report that human-resource specialists, because of their in-depth knowledge of the workforce, can assist in ensuring the effective utilization of those employees who lack a pre-designated disaster assignment in the overall response and recovery efforts. Along these lines, emergency managers have also indicated that human-resource officers can assist in disaster efforts through the provision of disaster-related training to employees. Such training efforts could possibly be incorporated into ongoing employee on-boarding and professional-development training regimens. Other important methods for human-resources personnel to assist in disaster-recovery efforts include maintaining awareness of the human-capital supply in the broader environment, supporting the provision of special assistance to employees affected by a disaster and coordinating efforts to increase employee retention and decrease absenteeism after a disaster. Similarly, human-resources personnel are critical to ensuring the local government navigates the policy and regulatory environment in its employment of disaster-related temporary workers and contractors.

In conclusion, government human-resources and human-capital management continues to evolve in the U.S. Disasters bring special stressors upon governments. Due to the central role played by a government's employees in its disaster-response and -recovery efforts, human-resources and human-capital specialists can bring unique perspectives to bear in emergency-planning and disaster-response efforts. In order to capitalize on their knowledge, human-resource managers must be included in the various networks of communication and relationship-building centered on disaster-preparedness activities. Pre-incident collaboration and coordination is the key to unlocking the potential of the relationship between human-capital management and emergency management. Taking advantage of the unique knowledge of the local-government workforce held by human-resource managers, emergency managers can align disaster-preparedness efforts with their organizations' long-term, large-scale human-capital development efforts in training, retention, and workforce planning. As most emergency-management efforts are limited through lack of staff, enhancing the relationship between emergency management and human-capital management would serve to strengthen the team responsible for local government's disaster-response and recovery activities.

Bibliography

Ballenstedt, Brittany. "Union Seeks to Halt Funding for DHS Personnel System." *Government Executive*, April 2, 2007, www.govexec.com/defense/2007/04/union-seeks-to-halt-funding-for-dhs-personnel-system/24101/.

Bea, Keith, Elaine Halchin, Henry Hogue, Frederick Kaiser, Natalie Love, Francis X. McCarthy, Shawn Reese, and Barbara Schwemle. *Federal Emergency Management Policy Changes After Hurricane Katrina: A Summary of Statutory Provisions. CRS Report for Congress, Order Code RL33729* (Congressional Research Service and the Library of Congress: 2006).

Box, Richard C., Gary S. Marshall, B. J. Reed, and Christine Reed. "New Public Management and Substantive Democracy." *Public Administration Review* 61, no. 5 (2001): 608–619.

Butler, David. "Focusing Events in the Early Twentieth Century: A Hurricane, Two Earthquakes, and a Pandemic," in *Emergency Management: The American Experience 1900–2010*, ed. Claire B. Rubin (Boca Raton: Taylor & Francis Group, 2012): 13–50.

California Government Code. *Sections 3100–3900*. Retrieved September 24, 2014 from www.leginfo.ca.gov/cgi-bin/displaycode?section=gov&group=03001-04000&file=3100-3109.

California Governor's Office of Emergency Services (OES) and State Compensation Insurance Fund. (2001). *Disaster Service Worker Volunteer Program (DSWVP) Guidance*. April 6, 2001. Retrieved September 24, 2014 from hazardmitigation.calema.ca.gov/docs/ESA-all8-06-final.pdf.

Cavanaugh, John C. "Effectively Managing Major Disasters." *The Psychologist-Manager Journal* 9, no 1 (2006): 3–11.

Coggburn, Jerrell D. "The Effects of Deregulation on State Government Personnel Administration." *Review of Public Personnel Administration* 20, no. 4 (2000): 24–40.

Dynes, Russell R. "The Functioning of Expanding Organizations in Community Disasters." University of Delaware Disaster Research Center Report Series #2. 1968. Retrieved January 3, 2014 from http://udspace.udel.edu/bitstream/handle/19716/1250/RS2.pdf?sequence=1: 1–86.

Edwards, Frances L. and Isabelle Afawubo. "Show Me the Money: Financial Recovery After Disaster." *The Public Manager* 37, no. 4 (2008): 85–90.

Federal Emergency Management Agency Emergency Management Institute. n.d. National Incident Management System courses. Retrieved January 3, 2014 from https://training.fema.gov/nims/

French, P. Edward and Doug Goodman. "An Assessment of the Current and Future State of Human Resource Management at the Local Government Level." *Review of Public Personnel Administration* 32, no. 1 (2012): 62–74.

Gabris, Gerald T., Keenan D. Drenell, and James Kaatz. "Reinventing Local Government Human Services Management: A Conceptual Analysis." *Public Administration Quarterly* 22, no. 1 (1998): 74–97.

Goodman, Doug and Stacey Mann. "Managing Public Human Resources Following Catastrophic Events: Mississippi's Local Governments' Experiences Post Hurricane Katrina." *Review of Public Personnel Administration* 28, no. 1 (2008): 3–19.

Goodman, Doug and Stacey Mann. "Reorganization or Political Smokescreen: The Incremental and Temporary Use of At-Will Employment in Mississippi State Government." *Public Personnel Management* 39, no. 3 (2010): 183–209.

Gruenfeld, Deborah H., Elizabeth A. Mannix, Katherine Y. Williams, and Margaret A. Neale. "Group Composition and Decision Making: How Member

Familiarity and Information Distribution Affect Process and Performance." *Organizational Behavior and Human Decision Processes* 67, no. 1 (1996): 1–15.

Hansen, Zeynep K., and Gary D. Libecap. "Small Farms, Externalities, and the Dust Bowl of the 1930s." *Journal of Political Economy* 112, no. 3 (2004): 665–694.

Homeland Security Act of 2002 (HSA). *Public Law 107-296, 107th Congress.* November 25, 2002.

Hy, Ronald John and William L. Waugh, Jr. "The Function of Emergency Management." In *Handbook of Emergency Management: Programs and Policies Dealing with Major Hazards and Disasters*, eds. William L. Waugh, Jr. and Ronald John Hy (Westport, Connecticut: Greenwood Press, 1990): 11–26.

Jacobsen, Willow, Jessica E. Sowa, and Kristina T. Lambright. "Do Human Resource Departments Act as Strategic Partners? Strategic Human Capital Management Adoption by County Governments." *Review of Public Personnel Administration* 34, no. 3 (October 2014): 289–301.

Kellough, J. Edward, and Lloyd G. Nigro. "Civil Service Reform in the United States: Patterns and Trends," In *Handbook of Human Resource Management in Government*, ed. Stephen E. Condrey (San Francisco: John Wiley & Sons, 2010): 73–94.

Lewis, Marianne W., Constantine Andriopoulos, and Wendy K. Smith. 2014. "Paradoxical Leadership to Enable Strategic Agility." *California Management Review* 56, no. 3 (2014): 58–77.

Lindell, Michael K. "Are Local Emergency Planning Committees Effective in Developing Community Disaster Preparedness?" *International Journal of Mass Emergencies and Disasters* 12, no. 2 (1994): 159–182.

Lynn, Dahlia and Donald E. Klingner. "Beyond Civil Service: The Politics of the Emergent Paradigms." In *Handbook of Human Resource Management in Government*, ed. Stephen E. Condrey (San Francisco: John Wiley & Sons, Inc., 2010): 45–71.

Mann, Stacey. Survey of Mid-Size Local Government Emergency Managers. 2013.

Monroe, Eason. "California's Crisis in Freedom." *Nation* 175, no. 17 (October 25, 1952): 377.

Moynihan, Donald P. "Homeland Security and the U.S. Public Management Policy Agenda." *Governance* 18, no. 2 (2005): 171–196.

Naff, Katherine C., Norma M. Riccucci, and Siegrun F. Freyss. *Personnel Management in Government: Politics and Process* (Boca Raton: CRC Press, 2013).

Nigro, Lloyd G., and J. Edward Kellough. "Personnel Reform in the States: A Look at Progress Fifteen Years after the Winter Commission." *Public Administration Review* 68 (November 2008): S50–S57.

Nigro, Lloyd G., and J. Edward Kellough. *The New Public Personnel Administration*. 7th edition. (Boston: Wadsworth Cengage Learning, 2014).

Perry, James L. "The Civil Service Reform Act of 1978: A 30-Year Perspective and a Look Ahead: Symposium Introduction." *Review of Public Personnel Administration* 28, no. 3 (2008): 200–204.

Rutzick, Karen. "OMB Not Deterred by DHS Personnel Reform Setback." *Government Executive*, August 17, 2005. Retrieved January 3, 2014 from www.govexec.com/federal-news/2005/08/omb-not-deterred-by-dhs-personnel-reform-setback/19914/.

Ryan, Richard W. "The Department of Homeland Security Challenges the Federal Civil Service System: Personnel Lessons From a Department's Emergence." *Public Administration & management: An Interactive Journal* 8, no. 3 (2003): 101–115.

Selden, Sally Coleman. "Human Resource Management in American Counties, 2002." *Public Personnel Management* 34, no. 1 (2005): 59–84.

Shiramizu, Sharilyn, and Amarjit Singh. "Influence of the National Performance Review on Supervisors in Government Organizations." *Leadership & Management in Engineering* 6, no. 4 (2006): 150–159.

Thompson, Fred, and Hugh T. Miller. "New Public Management and Bureaucracy versus Business Values and Bureaucracy." *Review of Public Personnel Administration* 23, no. 4 (2003): 328–343.

Thompson, F. J. *Classics of Public Personnel Policy*. 3rd edition (Belmont: Thomson Wadsworth, 2013).

Tompkins, Jonathan. "Strategic Human Resource Management in Government: Unresolved Issues." *Public Personnel Management*, 31, no. 1 (2002): 95–110.

United States Department of Agriculture Natural Resources Conservation Service. "75 Years Helping People Help the Land: A Brief History of NRCS." n.d. Retrieved September 29, 2014 from www.nrcs.usda.gov/wps/portal/nrcs/detail/national/about/history/?cid=nrcs143_021392

Section III
Finance

8 Major Disasters and Private Risk Financing

Pete Vloedman

Introduction

Major disasters are, by their nature, low-frequency and high-severity events. Since the 1980s, however, these events have had increasing levels of economic and humanitarian impact around the globe. When Hurricane Hugo made landfall near Charleston, South Carolina in 1989, it caused (at the time) the world's largest insured loss from a natural disaster, at approximately $3 billion (in 1990 dollars). Only three years later, Hurricane Andrew devastated south Florida and caused approximately $16 billion of insured losses (in 1993 dollars). Damage from natural disasters continues to grow with time, as does the gap between uninsured losses (representing the total economic impact of the disaster) and the amount of loss covered by insurance.

There are numerous reasons for the increase in economic impact from natural disasters. Population growth and increasing global urbanization are two key factors. In 1950, approximately 30 percent of the world's population lived in cities. By the year 2000, this number had grown to about 50 percent, or approximately three billion people. Projections by the United Nations show that by 2025 this figure will increase to 60 percent based on a global population estimate of 8.3 billion people. Many of these cities, particularly in Asia, are located in disaster-prone areas (Michele-Kerjan *et al.* 2011).

Increasing global wealth is another reason for the increased economic impact of natural disasters. Despite the global financial crisis in 2007 and 2008, worldwide household wealth in U.S. dollars rose by 112 percent from year-end 2000 to mid-2013. Some of this appreciation, however, was caused by the depreciation of the dollar against many major currencies. Still, even with exchange rates held constant, average net worth rose 48 percent for the period. In the U.S. the average net worth rose by 56 percent (Credit Suisse 2013). This increase in global wealth has resulted in a commensurate increase in assets, both private and public, that are exposed to natural disaster risk and will drive economic losses resulting from future catastrophes.

The Rising Cost of Natural Disasters in America

During the period 1980–2012, the frequency and severity of natural disasters in America has mirrored the increase seen in the rest of the world. Notably, there has been an increase in severe meteorological disasters, particularly tornadoes. Both the frequency and the severity of loss from tornado events have gone up significantly in the last decade. Of particular note was the year 2011, which saw the most severe tornadoes spawned since records have been kept (surpassing 1974) and, if losses from all tornadoes were aggregated, would rank as the fifth costliest natural disaster in U.S. history since 1900 (Guy Carpenter 2011). In terms of total economic loss, however, hurricanes have been the primary cause of loss from natural catastrophes in the U.S. since 1980, accruing approximately $146,132,280,000 across three decades. While the 2000's had the lowest number of hurricanes making landfall, those that did strike the U.S. caused significant losses (U.S. Government Accountability Office 2007). See Table 8.1.

While there is a hotly contested debate over the contribution of global climate change to the frequency and severity of atmospheric-based natural disasters, there exists a significant body of evidence, only touched on above, that there is an economic cost to be paid for failing to address issues related to global warming.

Since the mid-1990's the federal government's share of disaster expenditures has increased significantly, from just 32 percent in the year of Hurricane Andrew to 77 percent in 2008, the year that Hurricanes Gustav and Ike struck the Gulf coast (Kaplan 2014). This is likely due in part to private insurance companies' post-Hurricane Andrew exclusion of losses from hurricane storm surge flooding from personal and commercial property policies. As seen in the aftermath of 2012's Hurricane Sandy, these storm surge losses are now being covered by policies issued under the National Flood Insurance Plan (NFIP) or, in many cases, being completely

Table 8.1 Insured Losses Associated with Hurricanes by Saffir-Simpson, Category and Decade

	Categories 1,2		Categories 3, 4, 5		Total	
1980s	$807,422	(11)	$9,905,042	(6)	$10,712,464	(17)
1990s	$9,038,801	(11)	$29,099,303	(8)	$38,138,104	(19)
2000s	$8,071,619	(7)	$89,210,093	(7)	$97,281,712	(14)
Total	$17,917,842	(29)	$128,214,438	(21)	$146,132,280	(50)

Note: Totals do not include crop losses associated with hurricanes. Number of hurricanes associated with losses is included in parenthesis. Hurricane classification was based on peak intensity at landfall

Sources: GAO analysis of PCS and NFIP data; NOAA (hurricane intensity classification)

uninsured and ultimately falling to the federal government as the de facto insurer of last resort.

The Rise of Alternative Risk Transfer Products to Finance Natural Disaster Risk

The expansion of desktop computing power in the mid-1980s led to increased understanding of natural disaster risk by the insurance industry, particularly in the U.S. Deterministic and stochastic catastrophe simulation models took insurance company policy data and translated it into potential losses from earthquakes and hurricanes. The results shocked insurance managers and were ignored by some. Some models suggested that a single company could sustain losses approaching $1 billion from a single Florida hurricane. Executives likely wondered how it could be that the largest insured loss ever had just occurred from Hurricane Hugo in 1989, totaling $3 billion. In 1992 Hurricane Andrew proved the model results to be realistic, and suddenly the capital level of the global catastrophe insurance market seemed too small to handle the newly demonstrated models.

Work began in earnest to create financial instruments that could spread this large catastrophic loss potential beyond the $200 billion+ catastrophe insurance industry to the broader capital markets, where even a $150 billion hurricane loss would be less than the average daily volatility of the global investment market's trillions of dollars of capital. Many types of products were tried: contingent equity and surplus notes that allowed a company to access long-term equity or debt capital should a specific type of disaster occur, and futures options contracts on U.S. earthquakes and hurricanes traded on the Chicago Board of Trade that allowed real-time trading of disaster risk, were a few of the innovations that surfaced at the time.

The product that resonated the most with institutional investors was the catastrophe bond. It was a floating rate debt instrument that, like a corporate bond, paid the investor a periodic interest payment in return for the use of the investor's capital. There was a major difference from a regular corporate bond. In a typical corporate bond, the bond issuer is allowed to use the proceeds from the bond offering for general corporate purposes. In a catastrophe bond, the capital that is raised is held in a trust account and is only allowed to be released in specific circumstances, typically 1) when the bond matures and the capital is returned to investors or 2) when a disaster occurs that meets specific event criteria allowing the capital to be transferred to the bond issuer for the purpose of paying expenditures related to the subject catastrophe. Details regarding how a catastrophe bond is structured and used will be discussed in the MultiCat Mexico case study.

Investors liked this product because it paid them periodic investment income from a source where the risk of losing money (i.e. default risk, being the occurrence of a disaster) was uncorrelated to their other investments such as stocks or corporate/government debt. The catastrophe bond market, which began in 1994, has grown to over $25 billion in twenty years and has grown at a rate of over 20 percent per annum over the last decade.

Case Study: The Mexican Government-Sponsored MultiCat Mexico 2009 Catastrophe Bond

Background

In 2009 the Mexican government became the first sovereign nation to securitize a portion of its financial obligations arising from natural catastrophes using the World Bank's MultiCat catastrophe bond (hereafter called a "cat bond") program.[1] This section will discuss the rationale, use and impact of the cat bond to assist Mexico in funding post-disaster recovery efforts.

Mexico's Natural Hazards

Mexico is exposed to significant risk of catastrophic loss from various types of natural disasters, including hurricanes, earthquakes and volcanic eruptions. Located in the Inter-Tropical Convergence Zone with land borders on the Caribbean Sea, the Gulf of Mexico and the Pacific Ocean, the potential for damage from hurricanes is very high. While wind damage is the most obvious cause of loss, storm surge and flash flood-caused mudslides also present significant threats to life and property from the hurricane peril. This was illustrated in 1998 by Hurricane Mitch, which caused nine deaths and an estimated $1 million in damage in Mexico, and in October 2005 by Hurricane Wilma, which made landfall on the Yucatan Peninsula, in popular resort areas, disrupting the region's core tourism business and causing over $1 billion of insured losses (Michele-Kerjan *et al.* 2011).

Earthquake is also a major country risk in Mexico. Five major tectonic plates interact within Mexico's borders. Its location on the "ring of fire," where 80 percent of the world's seismic activity occurs, makes over 50 percent of the nation's territory susceptible to severe earthquakes, including the capital, Mexico City (International Bank for Reconstruction and Development/The World Bank 2012). These seismically active areas generate approximately 60 percent of the Mexican gross national product and contain a similar proportion of the population (G20: the Government of Mexco and the World Bank 2012). On average, more than ninety earthquakes of magnitude 4.0 or greater occur each year. The most costly earthquake in terms of economic damage and loss of life occurred on September 19, 1985, when a magnitude 8.1 temblor struck Mexico City. Economic losses from the event totaled over $11.4 billion with insured losses of $724 million (in 2013 dollars), making it the eighteenth most costly insured loss from an earthquake, according to Swiss Reinsurance Company (Pontoriero 2014). Approximately 100,000 homes were damaged or destroyed. The disaster killed approximately 9,500 people and injured over 30,000.

Disaster Risk Financing in Mexico

A month after the 1985 Mexico City earthquake, the Mexican government established the National Commission for Reconstruction. Its mandate was twofold: 1) to take steps to facilitate the reconstruction effort following the earthquake, and 2) to establish the organizations needed to assist citizens affected by future disasters. This resulted in the 1986 creation of the National System of Civil Protection (SINAPROC), which involved all levels of government as well as non-governmental organizations and the private sector. Coordination of this effort was the responsibility of the Ministry of the Interior. Its mandate included managing policies and mechanisms for pre-event hazard mitigation as well as post-event response and reconstruction (OECD 2008). The Ministry of Finance was tasked with the design of financial risk transfer mechanisms.

As in the U.S., the main source of funding for disaster recovery is the Mexican federal budget. Given its experience with devastating natural catastrophes in the past and the potential for more disasters to occur in the future, the Mexican government decided to be proactive in its approach to disaster risk financing. They desired to add an *ex ante* approach to their historical method of post-event disaster appropriations to minimize the immediate impact on public finances in the wake of a major catastrophe. This led to the creation in 1996 of the Natural Disasters Fund, also known as Fonden. Fonden is responsible for funding government expenditures resulting from a disaster that has already occurred. Its funds are used for urgent needs, such as the costs of search and rescue, shelter and other expenses incurred in the days and weeks immediately following a catastrophe. It also is used to fund reconstruction in the months and years following the event. A companion facility to Fonden, Fopreden (Natural Disaster Prevention Fund), is used to fund disaster prevention and mitigation efforts at the federal level. This tandem funding mechanism for federal disaster mitigation and recovery efforts will be discussed in more detail in the next section.

The government is required by law to provide a minimum level of funding for Fonden and Fopreden in each federal budget, ensuring a constant source of funding. Since 2006, federal budget law has mandated that at the beginning of each year the Fonden and Fopreden resources cannot be less than 0.4 percent of the total federal budget. While this ensures that money is always allocated for disaster mitigation and recovery purposes, it has also proved to be a limitation on government spending. As often happens in government budgets, the stated "minimum spending requirement" in practice becomes an "unofficial maximum" limit of spending (Topete, Rationale and Genesis of the Mexican Disaster Risk Strategy 2011). Government oil revenues are a contingent source of funding if budgeted funds are insufficient to pay for losses. In 1999 the Fonden Trust was created to hold the funds appropriated through the federal budget to ensure they were only used for their intended purposes. Each Mexican state also has a similar trust. Reconstruction costs in each state are financed 50 percent by the federal government through Fonden and 50 percent through the individual state's trust. Having these trusts in place before a disaster occurs allows for rapid deployment of capital post-event, as well as the efficient tracking of recovery expenditures.

The increasing frequency and severity of losses from natural disasters provided the Mexican authorities with motivation to look beyond funding Fonden using only federal government proceeds and led to changes in its rules of operation, specifically, allowing it to buy insurance and ultimately issue the MultiCat Mexico 2009 cat bond. There have been three years since the creation of Fonden in which the Fonden Trust reserve was insufficient to pay incurred catastrophic losses and additional contingent funding was required (Michele-Kerjan *et al.* 2011). The $290 million MultiCat Mexico 2009 cat bond issue served as another source of contingent funding for the trust, providing a cushion in the event that budgeted resources proved to be insufficient.

Using a Cat Bond Instead of Traditional Reinsurance

Because Fonden is used to fund expenditures in the immediate aftermath of a catastrophe, security of the funding source and speed of payment were considered essential characteristics for this contingent funding solution. In a traditional reinsurance contract, a reinsurance company makes a promise to pay the reinsured, subject to contractual terms, if a covered event occurs. The ability of the reinsurer to pay is typically judged based on its credit rating from firms such as Standard & Poor's and Moody's. In a cat bond, the funds to be paid in the event of a catastrophe are held in high-quality securities, such as U.S. Treasury bills, residing in a collateral trust fund. The collateral fund may only be removed from the trust upon 1) the occurrence of a covered catastrophe or 2) maturity of the bond. The ability to have funds "on standby" until they are needed gave the edge on capital security to the cat bond.

As for speed of payment, a traditional reinsurance contract is limited by the need to demonstrate "proof of loss" caused by the disaster to items in which the government had an insurable interest prior to the event occurring. The reinsurance contract would "indemnify," or make whole, the government for the value of that which was damaged. That is, the government would have to show that it had lost something of financial value due to damage from the disaster in order to collect on the reinsurance contract. As discussed previously, Fonden pays for costs associated with disaster recovery efforts such as search and rescue, temporary housing, food and other expenses. These costs, while directly related to the occurrence of the disaster, do not cleanly fit the definition of "loss" described above, as they did not exist prior to the occurrence of the disaster. These costs represent a contingent financing need on the part of Fonden, with the contingency being the occurrence of the natural disaster. For contingent financing needs such as these, the parametric funding trigger represents a distinct improvement over the traditional reinsurance indemnity trigger.

Parametric Trigger Design

As the name implies, a parametric trigger uses the physical *parameters* that describe the catastrophic event to activate the payment mechanism of the risk financing contract. The example described here, from the government's MultiCat Mexico 2009 cat bond issue, is used to illustrate the construction of such a trigger mechanism.

Three critical seismically active zones in Mexico were identified where additional funds would be needed beyond what were available in the Fonden Trust, and were labeled Earthquake Zones A, B and C. For each zone an analysis was done to determine how severe an earthquake would have to be for the government to mandate these additional funds. The critical parameters were the earthquake's intensity (the amount of energy released), represented by the unit Magnitude, and the depth below the earth's surface where the earthquake occurred. The intensity of the earthquake represents how much potential damage can be caused by the event, while the depth determines how much energy can be dissipated prior to the earthquake's shockwaves reaching the surface (similar to the theme of the fairy tale *The Princess and the Pea*, the greater the distance between the center of the earthquake and the earth's surface, the less the earthquake will be felt on the earth's surface). The organization chosen to provide the parameter values is important to both the bond issuer and investors. In this case, the Mexican government would like the values to be finalized as soon as possible to expedite payment of the bond proceeds. Investors in the bond want the information to be provided free of moral hazard, which is the risk of another party's manipulation of the parameters that could potentially harm the investors. To satisfy both of these needs it was decided that, for MultiCat Mexico 2009, the parametric triggers would be determined by non-Mexican government agencies; to satisfy this mandate, the U.S. Geological Survey was selected to establish earthquake parameters and the U.S. National Hurricane Center established hurricane parameters.

The resulting parametric triggers include a given magnitude and depth, specific to each of the three regions: the Northwest Cocos, the Central Cocos and Mexico City (Topete 2012). If these conditions were met by an earthquake occurring in one of the zones (in addition to the Mexican government declaring a state of emergency as a result of the event), 100 percent of the cat bond principal (US$140 million) would be paid to the Mexican government.

Launching MultiCat Mexico 2009

As the maturity date of its 2006 CatMex cat bond issue approached, the Mexican government hoped to expand upon its success by adding the hurricane hazard to the peril of earthquake hazard previously covered by CatMex. During the mid-2000s, the World Bank's Department of Treasury (World Bank Treasury) became active in developing mechanisms to provide disaster financing assistance to governments with underdeveloped domestic insurance industries. One of the mechanisms that it developed was the MultiCat program, a cat bond issuance platform that enables governments to use a standard legal, operational and documentation framework to obtain contingent disaster risk financing through the capital markets. By establishing the MultiCat program and partnering with global reinsurers, catastrophe modelers, law firms and investment bankers, the World Bank was uniquely positioned to coordinate the many aspects of a cat bond issue that could prove daunting to an individual sovereign attempting such an issue on its own, such as hazard identification and modeling and bond structuring and distribution. Mexico, although it benefited from a mature domestic insurance industry and had previous experience with a cat bond issue, saw the value in working with the World Bank to promote the global use of the MultiCat platform and agreed to be its first participant in the program, resulting in the MultiCat Mexico 2009 cat bond. For the MultiCat Mexico 2009 cat bond issue, the World Bank (in consultation with the Mexican government) chose Goldman Sachs, Swiss Re Capital Markets (Swiss Re) and Munich Re Capital Markets (Munich Re) as investment bankers for the transaction; AIR Worldwide as the catastrophe risk modeler; and Cadwalader, Wickersham & Taft as legal counsel (Topete 2012).

Cat Bond Structure

In the MultiCat program, the cat bond is issued by a Special Purpose Vehicle (SPV), whose sole purpose is to direct cash flows between the government sponsoring the bond and the bond's capital market investors. In the case of MultiCat Mexico 2009, the structure had to be modified to comply with Mexican law. Fonden obtained a parametric insurance contract through a local insurance company. That company's policy was 100 percent reinsured by Swiss Re, who acted as a pass-through between the local Mexican insurance company and the SPV.

The premium paid to cat bond investors is a function of the probability of the catastrophic event(s) occurring. Factors that contribute to calculating the probability are the type of catastrophic event (e.g. earthquake, hurricane), the size of the geographic area where the event will be covered by the bond and the time period that the bond covers (MultiCat Mexico 2009, three years). The World Bank engaged a third-party catastrophe modeling firm (AIR Worldwide) to perform an independent risk assessment that would be critical to establishing the loss probability for the bond issue. Using their proprietary catastrophe simulation models and data provided by the Mexican government, AIR Worldwide estimated the annual probability of losses to the bond issue. With these probabilities in hand, the investment bankers on the transaction (Goldman Sachs, Swiss Re and Munich Re), by surveying potential investors and using current cat bond market conditions, estimated the premium that investors would require to buy the cat bonds. After the unit cost of bond issues was determined, the Mexican government was able to determine the Original Principal Amount of each bond based on its post-event financial needs and its current ability to pay investors the Interest Spread for each class of bond. As previously stated, one of Mexico's goals in issuing MultiCat Mexico 2009 was to obtain funding for hurricane disaster response in addition to the earthquake funding provided by the previous CatMex cat bond. By canvassing potential investors, the investment bankers on the transaction determined that the most efficient bond structure would be to issue a separate bond for each area of hurricane concern, specifically: the Baja California Peninsula (Pacific hurricane; bond class B); the southwest coast (Pacific hurricane; bond class C); and the Yucatan Peninsula (Caribbean Sea hurricane; bond class D). The parametric triggers associated with these hurricane tranches required the storm track to pass through a specified box in the geographic area with a corresponding minimum central pressure (a measure of hurricane intensity) as is shown in Table 8.2 (Michele-Kerjan et al. 2011)

The details of the four MultiCat Mexico 2009 cat bond tranches are shown in Table 8.3.

Table 8.2 Bonds for Each Area of Hurricane Concern

Class B (Zone 1)	Class C (Zone 2)	Class D (Zone 3)
944 mb or lower	944 mb or lower	920 mb or lower

Source: Michele-Kerjan *et al.* (2011)

Table 8.3 MultiCat Mexico 2009 Note Classes Description

	Class A	Class B	Class C	Class D
Hazads Covered	Earthquakes in three regions	Hurricanes in zone 1	Hurricanes in zone 2	Hurricanes in zone 3
Original Principal Amount	$140,000,000	$50,000,000	$500,000,000	$500,000,000
Rating by S&P	B	B	B	BB-
Interest Spread (over U.S. Treasury money market funds)	11.50%	10.25%	10.25%	10.25%
Maturity	3 years	3 years	3 years	3 years

Source: Michele-Kerjan *et al.* (2011)

Issuance Timeline

In deciding when to launch a cat bond issue, consideration should be given to the hazards covered by the bond and any time dependencies embedded therein. In the case of MultiCat Mexico 2009, earthquake risk is not time-dependent (an earthquake can happen any day of the year), but hurricane risk is normally limited to the northern hemisphere tropical cyclone season, which runs from June 1 to November 30. Intense hurricanes that could potentially threaten the success of a bond issue, should they occur during the issuance process, typically occur during August, September and October. This led to the decision to launch the bond issue in October 2009.

The entire process of issuing the bond, from approval, to risk assessment and bond structuring, to investor marketing, took approximately nine months. Legal documentation preparation time was not a limiting factor since the bond issue took advantage of the standard "shelf" documents already prepared by the World Bank for the generic MultiCat program. For the most part, drafting was limited to the Offering Circular Supplement that contained terms specific to MultiCat Mexico 2009. The longest part of the process (four months) involved an iterative optimization exercise to determine the size of the geographic area which could be covered with a sufficiently high probability of payment to the government for a price that was within the project budget allocated by the Ministry of Finance (Michele-Kerjan et al. 2011). The marketing process was completed in less than a month, resulting in bond orders of 2.5 times the authorized issue size from a global cast of institutional investors (Michele-Kerjan et al. 2011). Clearly, the issue was an initial success (see Table 8.4).

POSTSCRIPT: HAS THE MULTICAT MEXICO CAT BOND SERIES WORKED IN PRACTICE?

Has the MultiCat Mexico cat bond program performed as the Mexican government planned? Since the first MultiCat Mexico bonds were issued in 2009 (subsequently reissued as MultiCat Mexico 2012 after the 2009 series matured), there have been four notable disasters from perils covered by the MultiCat Mexico cat bond. Of these four events, one may have triggered a payout on the bond.

In April 2010 an earthquake occurred on the Baja California peninsula, causing minimal damage in the cities of Mexicali and Calexico near the Mexico/U.S. border. The earthquake occurred far to the north of the earthquake covered areas for MultiCat Mexico and no bond payout was triggered. Later that year in July, Hurricane Alex made landfall near Soto la Marina, Tamaulipas. Some significant damage was reported, but no bond payout occurred as the storm's intensity was too low when it passed through the geographic area covered by the cat bond. It should be noted that when the MultiCat Mexico cat bond was reissued in 2012, the Pacific hurricane geographic trigger zone was expanded, possibly in an attempt to prevent a repeat of the Hurricane Alex scenario.

In September 2013 two hurricanes, Manuel and Ingrid, made landfall on the Pacific and Caribbean coasts (respectively) of Mexico within twenty-four hours of each other. These events caused significant flash flooding and mudslides and accounted for $5.7 billion in economic losses

Table 8.4 MultiCat Mexico 2009-I Notes—Investor Distribution by Investor Type

Investor Type	Amount Invested	Share
Specialist ILS Managers	$115.75 million	39.91%
Reinsurers	$97.50 million	33.62%
Bank	$25 million	8.62%
Hedge Fund	$21 million	7.24%
Money Managers	$14.75 million	5.10%
Reinsurer Cat Fund	$10.25 million	3.53%
Endowment/Penion Funds	$5.75 million	1.98%

Source: Michele-Kerjan et al. (2011)

and over $200 million of insured losses. Manuel caused about 85 percent of the total damage, with 123 fatalities, and displaced over 20,000 people. Moody's estimated that damage from the combined events would exceed the Fonden balance (Artemis.Com 2013). However, since neither of the storms made landfall in a geographic area covered by MultiCat Mexico, no bond payout occurred.

Finally, Hurricane Odile made landfall near Cabo San Lucas on the Baja California Peninsula on September 15, 2014, causing major damage to the city and significantly impacting the area's tourism industry. According to Moody's Investors Service, over 95 percent of local homeowners do not have insurance, meaning that the rebuilding costs will likely fall to state and federal facilities such as Fonden. Insured losses are estimated at $950 million (Carrier Management 2014). Odile made landfall in Hurricane Zone 1 of the MultiCat Mexico 2012 Class C cat bond at an estimated minimum central pressure of 930 millibars (Standard & Poor's Ratings Services 2014), within the 50 percent principal payout band for the cat bond (Michele-Kerjan *et al.* 2011). On September 18, 2014, Standard & Poor's Ratings Services placed the MultiCat Mexico 2012 Class C cat bond on CreditWatch Negative, based on publicly available information on Odile and the fact that Swiss Reinsurance Co. Ltd. had submitted an event notice to AIR Worldwide Corp., the Calculation Agent for the cat bond. While post-event meteorological data and trading activity indicated a market expectation of a 50 percent payout to the Mexican government in the months following the event, the final meteorological data provided by the National Hurricane Center resulted in no payout from the bond as a result of Hurricane Odile.

Disaster Risk Financing in America: Redesigning Our Process to Reflect Current Fiscal Reality

This section will discuss a concept to reduce America's reliance on federal budget-related post-disaster funding sources for both domestic and foreign disaster recovery assistance, as well as to provide a dedicated source of funds for domestic pre-event disaster mitigation programs. While there are many divisions of the federal government, and even more line items in the federal budget that play a role in disaster recovery, financing and mitigation, this discussion will focus on the following:

1 Funding for domestic disaster mitigation programs coordinated by FEMA (Federal Emergency Management Agency).
2 Domestic disaster relief funds provided by FEMA's Disaster Relief Fund.
3 Disaster relief to foreign governments coordinated through the U.S. Agency for International Development's (USAID) Office of U.S. Foreign Disaster Assistance (OFDA).

Background

In the U.S., state and local governments have traditionally been charged with responding to disasters. For years the federal government distanced itself from providing relief funding to recover from specific disasters. In 1887, President Grover Cleveland vetoed a Texas disaster relief bill, in part because

> I can find no warrant for such an appropriation in the Constitution, and I do not believe that the power and duty of the General Government ought to be extended to the relief of individual suffering which is in no manner properly related to the public service or benefit...
>
> (Wooster 2006)

President Theodore Roosevelt furthered this position in the wake of the 1906 San Francisco Earthquake when he named the American Red Cross the official agency to assist the city, describing it as "the only organization chartered and authorized by Congress to act at times of great national calamity" (The White House Office of the Press Secretary 2000). For a large part of U.S. history, federal disaster relief was provided on an *ad hoc* basis.

Although not specifically a disaster relief fund, the 1968 establishment of the National Flood Insurance Plan (NFIP) was the first permanent step taken by the federal government to address citizens' need for financial assistance arising from a major catastrophic peril. According to the NFIP, 90 percent of all national disasters in the U.S. have involved flooding. Because flooding is catastrophic in nature and difficult to predict, it is typically excluded from the standard private homeowners' insurance policy.

Two problems that have impacted the historic profitability of the NFIP are its low number of participants and inadequate premium rates. The Flood Disaster Protection Act of 1973 required the purchase of flood insurance in special flood hazard areas. Since then, homeowners in designated flood zones with mortgages held by federally regulated lenders have been required to buy flood insurance. Voluntary participation in the program is typically very low, with the exception of high-hazard areas. Complicating matters is that Congress has historically authorized FEMA to charge subsidized (actuarially inadequate) premium rates to encourage voluntary participation in the program. The result is that people that buy NFIP policies are more likely to experience a claim than is reflected by the low premium they are charged. The 2013 Biggert–Waters Act attempted to return rates to actuarially sound levels, but a 2014 revision has reinstituted some subsidies.

This high risk pool aspect is illustrated by the number of properties that submitted more than one NFIP claim from 1978 to 2011, designated "repetitive loss properties" by FEMA. Assuming each property represents one policy, when compared to the average number of policies during the period, of 3,482,257, and total claims paid, of $40,500,547,891, it can be seen that approximately 5 percent of the policies issued resulted in an astounding 30 percent of the total claims paid during the period (King 2013) (see Table 8.5).

The 1970s saw the federal government attempt to formalize its *ad hoc* disaster relief processes. Following the Flood Disaster Protection Act of 1973, President Nixon signed the Disaster Relief Act of 1974, which established a standard process for Presidential disaster declarations (Federal Emergency Management Agency 2014a). On April 1, 1979, President Carter created FEMA, an independent cabinet-level agency aimed at coordinating efforts to keep the nation safe from "acts of God." Its mission would shift at the height of the Cold War in the 1980s to ensuring continuity of government, and would not resume its focus on natural disasters until after Hurricane Andrew devastated Florida and Louisiana in 1992 (OECD 2008).

In the wake of the 9/11 attacks, the federal government shifted its focus from disaster preparedness to disaster response. It also shifted its threat priority from natural disasters to terrorism. This resulted in huge changes for FEMA. It lost its cabinet-level status and was placed under the newly created Department of Homeland Security (DHS). While this organizational change has had many adverse effects on FEMA's ability to perform its missions (as evidenced in 2005's Hurricane Katrina), it has caused several notable changes in financing procedures.

Prior to becoming part of the DHS, FEMA was responsible for approving and distributing all disaster mitigation grants. Post-9/11, a new organization was created within the DHS, named the Office of State and Local Government Coordination and Preparedness, that assumed responsibility for numerous grant programs, leaving FEMA financially

Table 8.5 Total Repetitive Flood Loss Properties in the NFIP: 1978–2011 (as of December 31, 2011: $ nominal)

Building Payments	$9,332,087,006
Contents Payments	$2,768,293,788
Total Payments	$12,100,980,774
Average Payment	$24,388
Number of Losses	496,178
Number of Properties	166,368

Source: U.S. Department of Homeland Security, Federal Emergency Management Agency

responsible for only three mitigation grant programs (the Hazard Mitigation Grant Program, National Pre-Disaster Mitigation Fund and the Flood Mitigation Assistance Program), and other programs such as the NFIP and the Disaster Relief Fund (Federal Emergency Management Agency 2014b).

FEMA's Hazard Mitigation Grant Program (HMGP)

The purpose of FEMA's Hazard Mitigation Grant Program (HMGP) is to ensure that the opportunity to take critical mitigation measures to lower the impact from future disasters is not lost during the reconstruction process following a disaster that has already occurred where a Presidential major disaster declaration has been issued. It provides for up to 15 percent of the first $2 billion of disaster assistance, 10 percent for amounts between $2 billion and $10 billion and 7.5 percent for amounts between $10 billion and $35.333 billion (Federal Emergency Management Agency 2013a). A key limitation of this program is that a disaster must have already occurred for the funding to become available. It is not available for mitigation efforts in areas where a hazard is known to exist (e.g. the New Madrid seismic zone earthquake risk) but a disaster has not yet occurred.

FEMA's National Pre-Disaster Mitigation Fund (PDMF)

FEMA's National Pre-Disaster Mitigation Fund (PDMF) was created in 1997 to assist state and local governments in implementing pre-disaster hazard mitigation measures. Unlike the HMGP, which draws its funds from the Disaster Relief Fund (DRF) and is part of the funds approved for a disaster that has already occurred, the PDMF is independent of the DRF and is not dependent upon a Presidential disaster declaration for funding (Federal Emergency Management Agency 2013b). This makes the PDMF much more flexible than the HMGP, in that mitigation efforts can be implemented in advance of a disaster occurring. FEMA proposed eliminating funding for the PDMF in FY 2013, citing overlap with other programs such as the Hazard Mitigation Grant Program (Federal Emergency Management Agency 2013b). However, Congress has continued to fund the program and increased its funding in FY 2014 to $63 million, from $23 million the previous fiscal year, returning funding to levels similar to recent years (Federal Emergency Management Agency 2014c) (see Table 8.6).

FEMA's Disaster Relief Fund (DRF)

FEMA's DRF is the primary federal funding source for disaster response and recovery. Presidential disaster declarations, issued in accordance with

Table 8.6 History of Pre-Disaster Mitigation (PDM) Appropriations: FY 1997 to FY 2009

Fiscal Year	Program	Amount Requested (in millions)	Appropriations (in millions)	
1997	Project Impact	N/A	$2	EMPA account
1998	Project Impact	$50	$30	EMPA account[a]
1999	Project Impact	$50	$25	EMPA account
2000	Project Impact	$30	$25	EMPA account
2001	Project Impact	$30	$25	EMPA account
2002	Project Impact	$–	$25	EMPA account
2003	PDM	$300	$150	PDM Fund established[b]
2004	PDM	$300	$150	PDM Fund
2005	PDM	$150	$100	PDM Fund[c]
2006	PDM	$150	$50	PDM Fund
2007	PDM	$100	$100	PDM Fund
2008	PDM	$75	$114	PDM Fund
2009	PDM	$75	$90	PDM Fund
2010	PDM	$150	Pending	

Notes: a. EMPA is the Emergency Management and Planning Assistance (EMPA) account, which is FEMA's general administrative account.
b. The separate PDM account creates a separate line item for PDM for the first time in the FEMA budget.
c. For the first time in legislative language P.L. 108-334 directed that the PDM funds "shall be awarded on a competitive basis."

Source: FEMA, Mitigation Directorate, June 2010.

the Robert T. Stafford Emergency Relief and Disaster Act, are necessary for DRF funds to be dispersed.

Disaster Relief Fund (DRF) Funding and Budgeting Practices

Congress funds the DRF annually through regular appropriations. A unique feature of these appropriations is that they never expire. Known as a "no-year account," any funds remaining at the end of the year are automatically carried over to the next fiscal year. A side benefit of this type of account, as recently illustrated by Congress, is that the funds remain available during a government shutdown or appropriation funding lapse (Tollestrup 2013). When DRF funds are near depletion, as has happened many times this century due either to extreme disasters—Hurricane Katrina (2005), Hurricane Sandy (2012)—or insufficient annual appropriations, Congress provides additional funding through supplemental appropriations (Lindsay 2014). FEMA is required to report details of the DRF to Congress on a monthly basis (see Table 8.7).

Historical shortfalls in the DRF have been attributed to two main factors: prior decisions to not budget for high-cost disasters in annual appropriations and the randomness of disasters occur over time (see Figure 8.1).

Funding for major disasters is based on FEMA spending for all past declared major disasters. Funding for non-major disasters uses a ten-year average of previous events. This new approach has resulted in higher DRF funding requests. For the FY 2000–2014 period, the average appropriation requested was $2.5 billion. The total average appropriation for the period, including supplemental appropriations, was $9.5 billion. Since enactment of the BCA, the requested appropriation for FY 2013 was $6.1 billion

Table 8.7 Requests, Appropriations and Supplemental Appropriations to the DRF: FY 2000–14

Fiscal Year	Administration Request	Enacted Appropriation	Total Supplemental	Appropriation
2000	$2,780	$2,780	$0	$2,780
2001	$2,909	$1,600	$2,000[a]	$3,600
2002	$1,369	$2,164	$7,008	$9,172
2003	$1,843	$800	$1,426	$2,226
2004	$1,956	$1,789	$2,500	$4,289
2005	$2,151	$2,042	$43,091	$45,133
2006	$2,140	$1,770	$6,000	$7,770
2007	$1,941	$1,487	$4,256	$5,743
2008	$1,652	$1,324[b]	$10,960	$12,284
2009	$1,900	$1,278	$0	$1,278
2010	$2,000	$1,600	$5,100	$6,700
2011	$1,950	$2,645	$0	$2,645
2012	$1,800	$7,100	$6,400	$13,500
2013	$6,089	$7,007	$11,485	$18,492
2014	$5,626	$6,220	N/A	$6,220

Notes: Does not include rescissions or transfers unless incorporated in appropriation acts. Does not include appropriations made in the same act to accounts other than the DRF.
  a. P.L. 107-38 appropriated $40 billion in response to the terrorist attacks of September 11, 2001. The legislation did not specify the amount to be allocated to the DRF, but required that not less than half must be allocated for disaster recovery and assistance associated with the airliner crashes in New York, Virginia and Pennsylvania. On September 21, 2001, President Bush notified Congress that $2 billion of the amount appropriated in P.L. 107-38 would be allocated to FEMA for disaster relief "in New York and other affected Jurisdictions."
  b. Does not include $2,900 million in FY 2008 emergency supplemental funding for Disaster Relief enacted by P.L. 110-116.

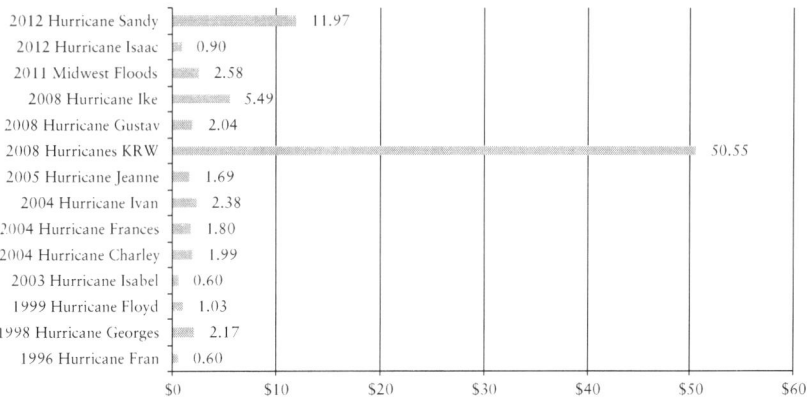

Figure 8.1 Disasters Costing FEMA $500 Million or More

($0.6 billion for non-major disasters and $5.5 billion for disaster relief cap adjustment), and for FY 2014 it was $6.2 billion ($0.6 billion for non-major disasters and $5.6 billion for disaster relief cap adjustment). This higher level of funding in recent years has reduced the need for supplemental appropriations. However, this funding mechanism will expire in FY 2021, after which the potential for annually underfunding the DRF will increase (see Figure 8.1).

Foreign Disaster Aid: USAID's Office of Foreign Disaster Assistance (OFDA)

In 1993, President Clinton designated the Administrator of the U.S. Agency for International Development (USAID) as the Special Coordinator for International Disaster Assistance (Committee on International Relations and Committee on Foreign Relations 2003). USAID's Office of Foreign Disaster Assistance (OFDA), established in 1964, is the primary agency for providing non-food humanitarian assistance funds to foreign governments. For the purpose of this analysis we will ignore Department of Defense disaster relief costs for operations such as those performed in Haiti, Pakistan and the Philippines in recent years. In FY 2012, the U.S. government contributed over $4 billion through USAID and the State Department to disaster relief efforts worldwide (Margesson 2015). Of these funds, approximately $600 million was for OFDA-sponsored efforts.

The President is authorized to borrow up to $50 million in any fiscal year from any section of the economic assistance portion of the foreign aid program if funding within the OFDA budget is inadequate. The borrowed funds are typically repaid by supplemental appropriation. While OFDA is capable of providing some direct assistance, most of its activities are carried out through grants to the United Nations and other aid organizations.

The Case for New Methods of Disaster Risk Financing

In the domestic and foreign disaster programs just discussed, U.S. government spending comes completely from annual and supplemental budget appropriations. Disasters are by their nature unpredictable, and therefore very hard to budget for on an annual basis. This is one reason why, in the global insurance industry, disaster insurance is a special subsector that has developed various financial instruments to deal with the realities of disaster risk financing in cost-effective ways. In difficult fiscal times such as those we face today, there is constant pressure across all levels of government to keep expenditures low and put off contingent costs from future events until the need actually arises. While it is tempting to "kick the can down the road" and ignore these contingent costs until they occur, there are a number of reasons to avoid this temptation.

A primary reason is simply that funding sources assumed to be available at all times may not be available when they are actually needed. An example of this potential can be drawn from September 2008. After Hurricane Andrew devastated south Florida in 1992, the state government established the Florida Hurricane Catastrophe Fund (FHCF) to ensure that all insurers in the state had access to a minimum level of catastrophe reinsurance in the event of another major hurricane landfall. Insurance companies operating in the state are required to buy a minimum level of coverage from the fund. The fund assets are made up of premiums previously paid and debt issued by the fund. The fund's outstanding liabilities that it might owe to policyholders after a major hurricane have typically been greater than its assets. The assumption has been that if the fund's liabilities due exceeded its assets, it would simply go to the bond market and issue more debt after the hurricane. That assumption probably would not have been possible in October 2008.

On September 13, 2008, Hurricane Ike made landfall near Galveston, Texas, causing approximately $12 billion of insured losses. Ike hit Texas the same weekend that Lehman Brothers filed for bankruptcy and the global credit crisis accelerated to a fever pitch. If Ike had hit Florida instead of Texas, and if it had caused more loss to the FHCF than it had assets to pay them, it is safe to say that it would have been difficult, if not impossible, for the FHCF to approach the bond market with a debt issue to raise enough capital in time to pay its claims. The result would have been the potential default of a state-sponsored insurer, and funding would have had to be provided by the State of Florida and possibly the federal government.

A secondary but still important reason is that the opportunity to reduce potential losses before a disaster occurs is lost. The value of risk mitigation efforts has been well documented. However, in a fiscal climate that promotes a triage mentality of not facing potential threats until events occur, moderate cost mitigation projects that could yield loss cost savings many times over are not pursued because more immediate issues require funding. If a source of government revenue was dedicated solely to funding disaster mitigation efforts at home, the need to kick the mitigation can down the road could be prevented.

A Proactive, *Ex Ante* Disaster Risk Financing Plan

This section sets out a proposal to revise the federal government's disaster risk financing process for both domestic and foreign disaster assistance, as well as domestic disaster risk mitigation programs, using the Mexican government disaster risk management framework as a template. The template is modified to minimize the impact on current U.S. government agencies and disaster financing programs. This proposal focuses on funding natural disaster expenditures. It is suitable for addressing terrorism risk financing as well, but that application is not explored here.

Disaster Risk Financing Structure

Recall from the MultiCat Mexico case study the ministerial responsibility for disaster risk management in the Mexican government. Applying this structure to the U.S., FEMA fulfills many of the same duties and responsibilities as the Mexican Ministry of the Interior for domestic disaster relief efforts, in that it applies funding to various pre- and post-disaster programs (such as the Hazard Mitigation Grant Program, PDMF and DRF). Currently, FEMA is responsible for obtaining funding for its disaster relief and mitigation programs, and these programs are funded solely through budget appropriations. In the Mexican government disaster risk-management model, the Ministry of Finance secures funding. For foreign disaster assistance the coordinating agency is USAID. In the U.S. government the comparable agency to fulfill this funding role would be the Department of the Treasury (Treasury).

In the new disaster risk financing model (the "new model"), FEMA would retain its role in coordinating domestic pre- and post-disaster programs. USAID's role in coordinating U.S. disaster relief efforts to foreign governments would remain unchanged. The major shift would be in the responsibility for obtaining the funds necessary to support these initiatives. Under the new model, the Treasury would have responsibility for obtaining both domestic and foreign disaster mitigation and relief funding. The Treasury is the logical agency to have responsibility for these purposes. As organizations whose mission is domestic disaster management (FEMA) and foreign disaster relief (USAID), FEMA and USAID do not possess sufficient knowledge of the world's capital and insurance markets to fully utilize all of the financial tools that are available today to fund its programs in the most cost-effective manner. The Treasury, through its relationships with major players in the global capital markets and organizations such as the World Bank, has the access and knowledge required to determine and obtain the most cost-effective approaches to funding the government's disaster financing programs. The Treasury's Office of Emergency Programs would be responsible for obtaining the necessary funds for both mitigation and disaster relief efforts, as the role appears to be in line with the office's responsibilities as outlined in Treasury Directive 23-01 (U.S. Department of Treasury 2012).

Domestic Disaster Relief Funds Provided by the FEMA Disaster Relief Fund (DRF)

FEMA's Disaster Relief Fund (DRF) functions very much like the National Disasters Fund (Fonden) in the Mexican model, and would retain its current structure and operational purpose in the new model. The main difference would be the method for obtaining funding. Annual budget appropriations for the DRF would continue, but steps would be taken to significantly reduce the need for supplemental appropriations.

The method for determining annual DRF appropriation requests employed by FEMA since the enactment of the Budgetary Control Act of 2011 (BCA) has appeared to meet DRF financing requirements except for the occurrence of major disasters. Since the ability to use this method will stop upon the expiration of the BCA in 2021, statues should be enacted to allow the continued use of the method after that date.

In reviewing DRF annual and supplemental appropriations since the year 2000, it is clear that the major source of funding volatility comes from supplemental appropriations required after a single major disaster, such as Hurricane Sandy in 2012, or a series of major disasters, such as Hurricanes Katrina, Rita and Wilma in 2005. Fortunately, financial tools exist that can be used to fund this volatility and reduce the need for supplemental DRF appropriations. In the new model, the Treasury would issue one or a series of cat bonds that would provide contingent funding to the DRF in the event of a single major disaster or a series of major disasters. Funds to pay the investor coupon required for the cat bond would be drawn from the DFR and be included in the annual DRF budget appropriation request.

FEMA would use its historical and hypothetical disaster data to work with the Treasury in determining the events to be covered by the cat bond, using as a basis for the analysis the disasters most likely to result in a need for supplemental appropriations. The Treasury could engage service providers such as investment banks, catastrophe modeling firms and legal counsel to assist in the cat bond structuring process, or it could partner with the World Bank to use capabilities already developed through the MultiCat program.

In order to minimize the time required for the funds to be transferred from the cat bond collateral trust to the DRF, parametric triggers (see "Parametric Trigger Design" in the case study above) would be used to cause a payout from the cat bond(s) to occur. For example, the trigger could be constructed such that a payout would occur if one or more Category 3, 4 or 5 hurricanes were to make landfall within 100 miles of certain U.S. cities such as New York, Norfolk, Miami or New Orleans. A similar method could be used for other disasters such as earthquakes and tornadoes.

The use of a cat bond or other similar contingent disaster risk financing product would provide a fully collateralized source of contingent financing for the DRF that would be available regardless of government shutdowns or other events that could prevent or slow additional necessary DRF funding after a major disaster.

Disaster Relief to Foreign Governments Coordinated through the U.S. Agency for International Development's (USAID) Office of U.S. Foreign Disaster Assistance (OFDA)

The largest change in process in the new model would be in providing foreign disaster assistance. The U.S. would work with the World Bank, United Nations and other organizations to expand the use of the World Bank MultiCat product to countries with underdeveloped domestic insurance industries or other special situations that make the use of the MultiCat product appropriate. Using a new Treasury vehicle called the Patriot Cat Bond Fund (PCBF), the Treasury would buy cat bonds issued by foreign governments or non-governmental organizations (NGOs) using funds that had previously been budgeted for disaster relief grants provided by USAID. Foreign aid through the purchase of cat bonds would constitute only part of the overall foreign disaster aid provided by the U.S. government. USAID would continue to receive funding through the current annual appropriation method for additional disaster relief needs.

For the affected foreign government or NGO, the final outcome from use of a cat bond would be the same: they would receive disaster funds for their use courtesy of the U.S. government, only this time funding would come from a cat bond that the Treasury has purchased instead of through a grant check from USAID.

The U.S. government, however, gains a significant benefit from using this new method. Previously, the federal government received nothing in exchange for its disaster relief grant but the gratitude of the foreign government. In buying the foreign government's cat bond in the new model, the federal government receives cash from the interest payment that the foreign government pays on its cat bond. That interest payment can range from approximately 5 to 15 percent of the investment made by the federal government when buying the cat bond. In the new model, this money would be redeployed by the federal government for domestic disaster mitigation use.

As noted above, in FY 2012 the U.S. government paid approximately $4 billion to foreign governments for disaster assistance. If $1 billion of this amount was invested in MultiCat-type cat bonds at an interest rate of 10 percent, $100 million would be generated for use in domestic disaster mitigation efforts.

Funding for Domestic Disaster Mitigation Programs Coordinated by FEMA and Other Agencies

As discussed above, there are numerous programs, such as the Hazard Mitigation Grant Program and the National Pre-Disaster Mitigation Fund, run by FEMA and other government agencies that provide funding for disaster mitigation efforts. These programs are funded through various line items in the budgets of the coordinating agencies, some as diverse as the Departments of Agriculture and Homeland Security. Since these programs have the same primary mission or core purpose of reducing loss from disasters, accountability and effectiveness would be significantly enhanced if the money to support these programs came from a single mitigation fund.

The proposed fund, which we will call the Disaster Mitigation Fund (DMF), would be similar to the National Disaster Prevention Fund (Fopreden) in the Mexican model. It would be administered by FEMA with funding provided by the Treasury, which would be responsible for raising the funds required to support domestic mitigation efforts throughout the government.

In the current U.S. model, mitigation expenditures are determined at the agency level. In the new model, the mandate of FEMA's Federal Insurance & Mitigation Administration (FIMA) would be expanded to include the coordination of all domestic disaster mitigation expenses government-wide. FIMA would prioritize the various agency mitigation programs and determine the level of funding to be provided to each program on an annual basis. If funds available from the DMF were deemed insufficient to support the required national mitigation expenditure, FIMA would be able to request additional budgetary funding from Congress.

Conclusion

Normal atmospheric, geological and hydrological patterns, exacerbated by climate change, demographics and migration, are pushing economic and insured losses from natural disasters higher and higher. Since the early 1990's advances have been made in disaster risk financing products, such as catastrophe bonds, that have not been used by the U.S. government to address its disaster risk financing needs.

After its experience in the 1985 Mexico City earthquake, the Mexican government has adopted a proactive approach to disaster risk financing. It has used mandatory budget funding levels for disaster mitigation and response efforts to build financing reserves for use in the event of a major disaster *before disaster strikes*. It has also drawn on the expertise residing in its Ministry of Finance to explore the most cost-effective methods for disaster risk financing. Part of this effort involved working with the World Bank and its MultiCat catastrophe bond program.

The current U.S. government disaster risk financing program draws only on budget appropriations for funding and ignores financing methods used by other governments and the private sector to mitigate financial risks. It also does not make use of the significant expertise in the global capital markets that resides in the Department of the Treasury. A new model for U.S. disaster risk financing is possible that combines the current method of budget appropriations with new risk financing products, such as catastrophe bonds, that can not only normalize government expenditures for domestic disaster recovery needs but also generate revenue that can be used to fund domestic disaster risk mitigation projects. In the era of increasing fiscal stress experienced by government today, serious consideration must be given to new and innovative risk financing methods that can reduce budgetary spending. Failing to incorporate these new risk funding measures will produce increases in government debt and deficit spending. Methods such as these can improve the safety and welfare of the U.S. public while at the same time providing fiscal benefits to the federal government.

Note

1 In 2006 the Mexican government, assisted by Swiss Reinsurance Company, issued the *CatMex* cat bond to back a reinsurance policy issued to the Mexican government to help fund its exposure to earthquake losses. Other earlier cat bonds such as Formosa Re (issued by Central Reinsurance Company) and Ianus (issued by Munich Reinsurance Company) covered government disaster losses but were not issued by the federal government itself, as with MultiCat Mexico.

Bibliography

Artemis.Com. "Recent Hurricanes in Mexico Show Potential for Broader Cat Bond Triggers." September 26, 2013. www.artemis.bm/blog/2013/09/26/recent-hurricanes-in-mexico-show-potential-for-broader-cat-bond-triggers/. Retrieved June 24, 2016.

Carrier Management. "Hurricane Odile Losses Mount for Mexican Insurers; Credit Negative, Moody's Says." 2014. www.carriermanagement.com/news/2014/09/22/129445.htm#.V21hT0s0Hs4.email. Retrieved June 24, 2016.

Committee on International Relations and Committee on Foreign Relations. *Legislation on Foreign Relations Through 2002, Section 493 Disaster Assistance Coordination*. Washington D.C.: U.S. Government Printing Office, 2003.

Carpenter, Guy. *Number of Natural Disasters in the United States from 1900–2015, by Type*. December 30, 2011. www.gccapitalideas.com/2011/12/30/2011-catastrophe-update-historical-global-losses-expensive-losses-in-the-united-states-tropical-cyclones-outlook-for-2012/. Retrieved June 24, 2016.

Credit Suisse. *Global Wealth Databook 2013*. October 2013. https://publications.credithttps://publications.credit-suisse.com/tasks/render/file/?fileID=1949208D-E59A-F2D9-6D0361266E44A2F8-suisse.com/tasks/render/file/?fileID=BCDB1364-A105-0560-1332EC9100FF5C83. Retrieved June 24, 2016.

Federal Emergency Management Agency. "Hazard Mitigation Assistance Unified Guidance." 2013a. www.fema.gov/media-library-data/15463cb34a2267a900bde4774c3f42e4/FINAL_Guidance_081213_508.pdf. Retrieved June 24, 2016.

Federal Emergency Management Agency. "National Pre-Disaster Mitigation Program: Congressional Justification." 2013b. www.fema.gov/pdf/about/budget/11d_fema_national_pre_disaster_mit_fund_dhs_fy13_cj.pdf. Retrieved June 24, 2016.

Federal Emergency Management Agency. *About the Agency*. 2014a. www.fema.gov/about-agency. Retrieved June 24, 2016.

Federal Emergency Management Agency. *Grants*. 2014b. www.fema.gov/grants. Retrieved June 24, 2016.

Federal Emergency Management Agency. *Pre-Disaster Mitigation Grant Program*. 2014c. www.fema.gov/pre-disaster-mitigation-grant-program. Retrieved June 24, 2016.

G20: the Government of Mexico and the World Bank. "Disaster Risk Management in Mexico: From Response to Risk Transfer." In *Improving the Assessment of Disaster Risks to Strengthen Financial Resilience*, 211–222. 2012. https://issuu.com/world.bank.publications/docs/gfdrr_g20_high?backgroundColor=. Retrieved June 24, 2016.

International Bank for Reconstruction and Development/The World Bank. "FONDEN Mexico's Natural Disaster Fund—A Review." 2012. www-wds.worldbank.org/external/default/WDSContentServer/WDSP/IB/2015/07/10/090224b0828b5839/1_0/Rendered/INDEX/FONDEN000Mexic0ster0fund000a0review.txt. Retrieved June 24, 2016.

Kaplan, Alex. "Public–Private Partnerships in Risk Management." March 8, 2014. https://ercwts.files.wordpress.com/2014/03/2-kaplan_csg_wts.pdf. Retrieved June 24, 2016.

King, Rawle O. *The National Flood Insurance Program: Status and Remaining Issues for Congress*. Cogressional Research Service, 2013. www.fas.org/sgp/crs/misc/R42850.pdf. Retrieved June 24, 2016.

Lindsay, Bruce R. *FEMA's Disaster Relief Fund: Overview and Selected Issues*. Congressional Research Service, 2014. www.fas.org/sgp/crs/homesec/R43537.pdf. Retrieved June 24, 2016.

Margesson, Rhoda. *International Crises and Disasters: U.S. Humanitarian Assistance Response Mechanisms*. Congressional Research Service, 2015. www.fas.org/sgp/crs/row/RL33769.pdf. Retrieved June 24, 2016.

Michele-Kerjan, Erwann, Ivan Zelenko, Victor Cardenas, and Daniel Turgel. "Catastrophe Financing: Learning from the 2009-2012 MultiCat Program in Mexico." *OECD Working Papers on Finance, Insurance and Private Pensions*, No. 9. OECD Publishing, 2011.

OECD. "Financial Management of Large-Scale Catastrophes." *Policy Issues in Insurance* 12. Paris: OECD Publishing, 2008.

OECD. *Policy Issues in Insurance: Financial Management of Large-Scale Catastrophes*. Paris: OECD, 2008.

Pontoriero, Caterina. *20 Costliest Earthquakes by Insured Losses*. April 23, 2014. www.propertycasualty360.com/2014/04/23/20-costliest-earthquakes-by-insured-losses. Retrieved June 24, 2016.

Standard & Poor's Ratings Services. "Rating on MultiCat Mexico Ltd.'s Series 2012-I Class C Notes on CreditWatch Negative After Hurricane Odile Made Landfall." September 18, 2014.

The White House Office of the Press Secretary. "American Red Cross Month, 2000—A Proclamation." February 29, 2000. https://clinton4.nara.gov/WH/New/html/20000301_2.html. Retrieved June 24, 2016.

Tollestrup, Jessica. *Federal Funding Gaps: A Brief Overview*. Congressional Research Service, 2013. www.fas.org/sgp/crs/misc/RS20348.pdf. Retrieved June 24, 2016.

Topete, Rubem Hofliger. "Rationale and Genesis of the Mexican Disaster Risk Strategy." *High Level Roundtable on the Financial Management of Earthquakes*. Paris, 2011.

Topete, Rubem Hofliger. "The Mexican Disaster Risk Rinancing Strategy." 2012. www.lloyds.com/~/media/images/the%20market/communications/events/mexico/mexico%20week/rubem%20hofliger%20topete%20%20the%20mexican%20disaster%20risk%20financing%20strategy.pdf. Retrieved June 24, 2016.

U.S. Department of Treasury. "Treasury Directive 23-01." 2012.

U.S. Government Accountability Office. "Climate Change: Financial Risks to Federal and Private Insurers in Coming Decades Are Potentially Significant." 2007. www.gao.gov/products/GAO-07-285. Retrieved June 24, 2016.

Wooster, Martin Morese. "America's Response to Disaster: The Changing Role of the American Red Cross." *Capital Research Center*. August 1, 2006. https://capitalresearch.org/2006/08/americas-response-to-disaster-the-changing-role-of-the-american-red-cross/. Retrieved June 24, 2016.

World Bank Treasury. "MultiCat Program Product Note." April 26, 2011.

9 Financial Resiliency by Local Governments to Natural Disasters

Robert Bland, Jesseca E. Short, and Simon A. Andrew

Introduction

The social and behavioral sciences have recognized the importance of resiliency of an individual, community or organization to bounce back from external shocks or changes in their environment. For example, the ecological literature uses "resiliency" to mean systems' ability to function and to absorb exogenous disturbances while maintaining the same relationship among populations (Holling 1973; Walker et al. 2004). In psychology, resiliency means "maintaining good functioning after exposure to stress," which implies organizations' ability to cope with adversity and retain control while positively accepting change and securing good relationships (Bonanno 2004; Bonanno et al. 2005; Fergus and Zimmerman 2005; Campbell-Sills and Stein 2007). In a public-administration context, resiliency broadly implies the capacity of organizations to adapt to, recover from or accept change.

This chapter examines the financial resiliency of local governments in the U.S., and specifically their capacity to recover economically and socially from disasters and other hazards occurring as a result of natural forces or technological mishaps, or a combination of the two. Financial resiliency aims to reestablish financial stability and sustainable economic growth within a political jurisdiction as a result of a hazardous event. At the most general level, financial resiliency refers to local-government administrators who pursue innovative, flexible and creative strategies that restore economic and social normalcy after the consequences of natural and technical disasters.

This chapter is guided by three questions: 1) What is financial resiliency? 2) What financial indicators best capture the concept of financial resiliency? 3) How can local managers utilize these indicators to promote financial resiliency?

The chapter first examines the concept of resiliency, before discussing the meaning of financial resiliency and the indicators that local administrators can use to monitor and mitigate the negative consequences of disasters. These questions are important because they provide insights and guidance to practitioners on strategies that they can adopt in order to mitigate the consequences of disasters. Moreover, they highlight the types of strategies that can be used to increase local governments' ability to identify sources of financial resiliency and establish innovative ways to incorporate risk in their budgets.

What is Resiliency?

Disasters provide scholars in public administration with an opportunity to study behavior at the margin and, of equal importance, how policies affect organizations' capacity to mitigate the effects of disasters. From a public-finance perspective, disasters offer an opportunity to better understand what policies and practices lead some local governments to bounce back from disasters more quickly than others. In the formative period for public-management theory, Fayol (1918) maintained that the acceptance of change depended on having the right span of control (i.e. the optimum number of employees reporting to an administrator) and hierarchy. More recently, Pfeiffer and Salancik (1978) hold that organizational capacity to adapt to change depends not just on the span of control or having a clear set of rules and procedures, but also on an organization's ability to acquire and maintain resources.

In disaster management, resiliency is defined as the capacity to cope with and improvise in response to changing and ambiguous situations (Wick *et al.* 1999). Resiliency in disaster situations requires careful planning and improvisation. Disaster management requires dealing with multiple single entities, such as volunteers, other local governments, non-profit organizations and state and federal governments, under complex and unpredictable conditions (Kendra and Wachtendorf 2003).

From a psychological perspective, resiliency involves "maintaining good functioning after exposure to stress" (Campbell-Sills and Stein 2007). It is the ability to cope with adversity (Bergeman *et al.* 2006; Bonanno 2004; Bonanno *et al.* 2005; Fergus and Zimmerman, 2005). By understanding resiliency at an individual level, psychologists can help victims cope with psychological distress and recover from inevitable factors such as job loss, divorce, financial issues, natural disasters and the loss of a loved one. Bonanno (2004) notes that resilience is different from recovery. Resiliency maintains equilibrium across time, whereas recovery has points that fall below the threshold for normal activity. Resilient individuals, organizations or communities have the ability and capacity to become innovative and adapt to changing conditions. Resiliency follows a more stable pattern in the return to normalcy. Recovery is less flexible and less prepared for the unexpected. Recovery is characterized by a less stable trend and greater uncertainty in the quest for normalcy.

Economics defines resiliency as the "return to a fixed and narrowly defined equilibrium" (Christopherson *et al.* 2010, p. 3). However, there are multiple equilibriums—economic, political and regional. In economics, resilience is viewed in time, space and process (Christopherson *et al.* 2010). Time is measured by pre-shock, shock and post-shock. Space is constructed by human action (i.e. geographic and political boundaries). Process is political or economic (i.e. the decision to invest or disinvest in one region over another region). Community resilience is the ability of a community to sustain itself and recover from adversity (Chandra *et al.* 2011), with the aim of returning to normalcy. The importance of studying resiliency at the community level is to create stability and economic growth. Communities that bounce back more slowly from economic downturns or natural disasters tend to have had declining population and business activity prior to the disaster.

Pike *et al.* (2010) expand on this definition, making the distinction between resilience and adaptation, which is "short-term" recovery. For Pike *et al.* (2010), a resilient region maintains its capacity to recover from external shocks. Adaptability, for example, encourages local-government officials to cope with unforeseen or unexpected conditions. Local-government administrators must be risk-takers in order to abandon commonly used practices that have proven successful over time and establish different strategies, which encourages greater flexibility to manage unforeseen or unexpected conditions to achieve long-term benefits. Meanwhile, adaptation only handles the situation at hand. It responds to and takes actions that provide short-term benefits.

What is Financial Resiliency?

Financial resiliency has been defined from both an individual and an organizational perspective. At the household level, Jacobsen *et al.* (2009) define financial resiliency as the ways in which people access, build and preserve their financial assets and limit their liability. From the organizational perspective, Gulati (2013) defines financial resiliency as the availability of resources for response, recovery and reconstruction that does not disrupt daily fiscal activities.

In the context of emergency management, natural disasters can create short-term and long-term setbacks to development through the loss of infrastructure, lives and business activity as a result of diverting funds from routine activities and toward recovery and reconstruction. For example, the disaster-relief fund administered by the Federal Emergency Management Association (FEMA) dropped below $1 billion following the diversion of $7 billion to respond to the destruction caused by Hurricane Irene in August 2011 (Pittman 2011). FEMA stopped accepting and approving applications to fund repairs and reconstruction in places like

Joplin, Missouri, which had been struck by tornadoes and Cedar Rapids, Iowa, which was recovering from flood damage.

After any disaster that exceeds a local government's financial capacity for recovery, that government must look to state and federal sources for financial relief. However, the political shift rightward that has occurred in the U.S. and the growing public resistance to recognizing the collateral effects of disasters has raised the financial stakes for state and local governments. The political furor over funding relief to northeastern states following Hurricane Sandy illustrates the growing pressure for states and local governments to assume greater financial responsibility for disaster response and recovery. That pressure also reinforces the need for local governments to pursue policies that facilitate their financial resiliency in the event of a disaster.

The political shift rightward poses a new challenge for state and local governments in areas that are prone to disaster, such as earthquakes in California, hurricanes along the Gulf Coast, droughts in the southwest and wildfires on the urban fringe. If the collateral damage from a major disaster is increasingly regarded as a state and local responsibility, then what incentive do households and businesses have to invest in disaster-prone regions where the cost of living and doing business will inevitably be higher than is the case in less disaster-prone regions? If, as a result of global warming, disaster-prone areas face increasing losses from more frequent and intense events, at what point does it become economically inefficient to make further investment in protecting, not to mention restoring, a region from the effects of disasters? Should such regions of the country be allowed to decline economically, much like the sinking of Venice, Italy?

The changing politics of national disaster relief also elevate the importance of local governments taking action to increase their financial resiliency. The current formula for federal funding appears to reward those state and local governments that make the least effort to adopt policies that promote financial resiliency. Should federal (or even state) aid for disaster response and recovery be linked to a local government's investment in resiliency? How can federal and state policies be modified to create incentives for local governments to take greater responsibility for their own destiny, especially in regions with a higher incidence of natural disasters?

Indicators of Financial Resiliency

The purpose of financial resiliency is to mitigate fiscal pressures on federal, state and local governments. We use ten indicators to gauge the financial resiliency of a local government: (1) job market shifts; (2) population trends; (3) market value of property; (4) per capita income; (5) local revenues and expenditures; (6) property and casualty insurance; (7) local debt; (8) unfunded liability/fund balance; (9) long-term investment in infrastructure; and (10) management. The following sections review these indicators and their utility in gauging a local government's capacity for financial resiliency to disasters.

Local governments cannot rely solely on state and federal governments to help them fiscally prepare for natural disasters. Kloha *et al.* (2005) conducted a fifty-state survey that found that although "oversight of local fiscal behavior is a primary responsibility for states, it is not being carried out diligently and effectively in most states" (p. 252). Only fifteen states used indicators to evaluate their local governments' preparation for disasters. Seven of those states used both early warning and ex post declarations of fiscal distress. As for reliance on federal financial aid, the criteria for local governments to receive aid is subjective, typically based on the geographic region of the country and the magnitude of the natural disaster.

This is important because state, and especially federal, governments lack the resources to collect data and monitor the fiscal condition of local governments. Only after a disaster hits and its economic impact on a region is fully assessed do these overlapping governments become aware of local governments' financial vulnerability. As such, the front line of defense for financial resiliency is at the local-government level.

Job Market Shifts

Shifts in the distribution of employment across sectors provide a key indicator of financial resiliency. First, employment in higher-waged sectors provides jobs for individuals, which influences their purchasing power and, in forty-one states, a portion of their revenue from state income taxes. Second, sales tax from the provision of goods and services contributes a significant stream of revenue in forty-five states. Last, the employment sector is an indicator of financial resiliency because jobs bring people to reside in cities or towns. An increase in population increases the tax bases used by local governments, enabling them to invest in more resilient equipment, facilities, technology and infrastructure to offset the costs of potential damages caused by natural disasters.

Although disasters disrupt or stifle economic activity, studies have found that some markets had greater economic activity after the disaster than before (Ewing and Kruse 2002; Ewing *et al.* 2003; Guimaraes *et al.* 1993; Skidmore and Toya 2002). Gillespie (1991), for example, found that although Hurricane Hugo had a negative impact on employment, it was short-lived. In the long term, the construction industry added over 8,000 jobs by the spring to build, repair and reconstruct buildings from the damages caused by the hurricane, which offset the loss of jobs in the tourism and trade sectors. This example indicates that it is not the number of businesses in a location, but the composition of the economic base that affects how quickly a community can bounce back from a disaster.

Population Trends

Population growth in various locations, such as the coastal region, has increased the number of people and properties at risk. The more people in disaster-prone areas, the higher the risk of loss of property and lives due to natural disasters. Furthermore, regions that are disaster-prone likely draw and retain more vulnerable populations as a result of a lower cost of living and lower mobility to move.

Such areas likely have relatively higher rates of unemployment as a result of the higher cost of production, further adding to the vulnerability of at-risk populations. The higher cost of production is particularly apparent for more capital-intensive industries, such as manufacturing and transportation. These sectors face higher insurance rates in disaster-prone regions because of the greater cost of replacing their facilities damaged or destroyed by a disaster. Adding to their production cost may be the higher cost of construction from local building codes that require more disaster-resistant structures. Finally, pre-disaster trends in unemployment and business disinvestment are exacerbated by disasters, adding to the cost of recovery and restoration to pre-disaster levels of economic growth.

Market Value

Areas with higher levels of development—residential or commercial—tend to have greater losses as a result of their higher property values. The financial impact of a disaster may be lowered with better preparedness, such as building construction codes and advanced weather and seismic monitoring devices. Although development is encouraged in order to increase the local tax base, thereby increasing local revenues, development also increases the costliness of a disaster. According to the National Oceanic and Atmospheric Administration (NOAA), 39 percent of the U.S. population is concentrated in counties directly on the shoreline.

One consequence of the migration of high-income households to coastal areas is that it pushes poorer households to regions more poorly prepared for disasters (Zedlewski 2006). In 2006, Congress removed FEMA's responsibility for housing (Alpert 2006). Yet, months later, the federal government provided millions of dollars to Louisiana and Mississippi for temporary storm shelters until owners could rebuild. The increase in displaced families only added to the queue of those waiting for public assistance vouchers for rent.

Per Capita Income

Wealth is another important indicator of financial resiliency for local governments. More wealthy communities possess greater resources to recover more quickly from a disaster. Cutter *et al.* (2003) used county-level socio-economic and demographic information to construct an index for social vulnerability. With regard to per capita income, they found that wealth accounted for 12.4 percent of the variance. Higher wealth is also associated with better technology, disaster-resistant infrastructure and facilities, which mitigate costly damages from natural disasters. Toya and Skidmore (2007) found that countries with higher per capita income experienced less damage and loss of life.

Long-Term Investment in Infrastructure

Infrastructure includes, but is not limited to, electric power, water, transportation, communication networks, storm drainage and public buildings and structures. According to the *New York Times*, Hurricane Sandy cost $33 billion in repairs to damaged housing and infrastructure and $9 billion to help protect transit systems, the power network and sewage treatment facilities from future storms. Delays in repairing and rebuilding infrastructure have both direct and indirect costs (e.g. costly adaptation or utility reduction from loss of use).

Chang (2003) conducted a case study in Portland, Oregon using an extended life-cycle cost analysis framework to argue that overhead costs should be considered along with the cost of deterioration over time. Life-cycle cost analysis is similar to cost–benefit analysis but has the advantage of including changes over time, such as infrastructure deterioration, long-term maintenance costs and urban growth. When local-government administrators make decisions with regard to infrastructure, revenue loss must also be accounted for.

Chang (2003) tests three mitigation options—no seismic retrofit, pipe replacement and tank and pump upgrades—to identify the best approach to handle seismic risks. In this study, the author identifies societal and agency costs. For example, pump upgrades were found not to be cost-effective for the agency, but cost-effective for the society contrary to no retrofit. However, pipe-replacements were the best option for both the society and the agency, but the resources needed to pursue this option ($378 million) were much greater than the financial benefits for the agency or the society. Policymakers should be concerned, because agencies may choose the option that is solely in their best interest.

Local Revenues and Expenditures

Although intergovernmental transfers served as a major source of finance to local governments (Wildasin 2009), budget cuts at the federal and state levels have constrained the amount of aid flowing to local governments. Unexpected occurrences such as natural disasters can severely increase expenditures, while revenues remain stagnant or even decline in the short term.

Benson and Clay (2004) find that natural disasters hinder long-term development because governments have an obligation to provide goods and services for immediate needs after a disaster, which takes away investment in new improvements. There may be an imbalance of revenues and expenditures. Governments may not have generated enough revenue to cover the expenditures necessary to provide services to residents. In addition, local governments may not have been able to acquire capital from the bond market because of their losses in the tax base following a disaster. In place of increasing taxes and reducing expenditures in order to close the gap, local governments often have no recourse but to look to state and federal government for financial assistance to restore their economy.

Property and Casualty Insurance

Property and casualty insurance offers another tool that mitigates financial stress for local governments. Property insurance provides a critical safety net for homeowners and businesses in financing the cost of disaster-related damages. Insurance companies have become highly regulated with regard to requesting stricter building codes and granting insurance to those entities that meet the code requirements. Financial institutions could require an inspection stating the condition of homes or businesses for the purposes of obtaining a loan or mortgage (Kunreuther 1996).

Some states may offer supplemental insurance plans for more disaster-prone areas. For example, the Texas Windstorm Insurance Association (TWIA), created in 1971 by the legislature following Hurricane Celia, provides property insurance as a last resort to homeowners and businesses in fourteen coastal counties in the state. The state program only insures against losses caused by wind and hail and is available only if no private insurance is available to the property-owner in designated high-risk areas along the seacoast. The effect is to shift to the state the highest-risk properties in the most hurricane-vulnerable areas of the state.

Local Debt

Local governments typically use long-term debt to finance capital improvements. In the case of General Obligation bonds, the debt is repaid through an annual levy on taxes; in the case of Revenue bonds, it is repaid through service charges collected from utility customers. More severe disasters may interrupt such annual payments as a result of lost property value or utility capacity, and with it a greater potential for default.

Because of the greater risk of default, it is likely that bond ratings for state and local governments in disaster-prone areas are lower, thereby increasing their cost of borrowing. Here, again, the cost of doing business for both governments and businesses may be higher in these areas, thereby raising the cost of living or doing business.

Fowles *et al.* (2009) use a sample of 513 California municipal bonds issued between 2004 and 2006 to identify the relationship between the municipal bond market and earthquake risk. The authors find that issuing debt after Hurricane Katrina paid a premium to investors that was proportional to the municipality's assessed underlying earthquake risk. In other words, earthquake risk became a significant predictor of borrower interest costs only after Hurricane Katrina. More definitive research is needed to empirically assess the long-term effect of disasters on the cost of debt for state and local governments, but it is reasonable to conjecture that the uncertainty raises borrowing costs.

Budget Reserves and Contingency Funds

There is little research evaluating the reserved funds used by local governments in cases of emergencies such as natural or other types of disasters, economic recessions or other interruptions to normal economic activity. Fund balance refers to the difference between assets and liabilities reported in a governmental fund. Monies set aside in a budget-stabilization fund, typically reported as part of the fund balance, can provide a financial cushion that helps preserve the continued orderly operations of government and the provision of services. Governments, like businesses and households, need a financial cushion against the potential shocks of unanticipated circumstances and events such as revenue shortfalls and disasters.

While the federal level does not mandate state and local governments to establish rainy-day or budget-stabilization funds, there are existing financial strategies utilized by local governments which enhance financial resiliency. The state of Texas, however, has a disaster-contingency fund. Currently, the funding for this disaster-contingency fund is being debated in the 81st Legislature. The city of Coppell, Texas, for example, has a reserve fund established in its budget that amounts to at least 10 percent of the proposed expenditures for the major operating funds stated in its city charter Section 7.03 Budget. Harris County, Texas, has a Public Improvement Contingency Fund in Section 6 of its investment policy that is used to assist with capital projects and unforeseen catastrophic events.

Management

With external shocks such as economic downturns, local governments incur reductions in revenues that inevitably cause reductions in public spending. As a result, managers utilize a conservative approach when preparing their revenue forecasts to ensure that day-to-day expenditures are covered. Revenue forecasts depend on the stability of trends in prior periods and the number of periods available. They cannot predict shocks and irregularities caused by external factors such as disasters or economic recessions.

A local government's ability to bounce back requires that it take reasonable precautions for the uncertainty that it faces. Disaster-prone areas have an added level of uncertainty in the management of their financial resources that requires managers to be proactive in scanning their budgets and services to identify their level of vulnerability and strategies to reduce their financial exposure.

How Can Local Managers Utilize these Indicators to Promote Financial Resiliency?

Local governments can financially prepare for natural disasters by utilizing the indicators of financial resiliency. There are several factors that local governments must consider if they aim to be financially resilient. First, a proven strategy is collaborating with other governments, nonprofits and for-profit organizations to share costs (McGuire and Scheck 2010). Contracting-out services and engagement in interposal agreements/mutual aids are strategies that require information and negotiation from all parties to reduce fiscal stress and ultimately achieve the goal of providing efficient goods and services to residents immediately after disasters.

Also, local governments must be cognizant of population growth and per capita income in areas that are vulnerable to natural disasters. Although local governments view population growth as an increase in their revenue stream, in the wake of a natural disaster it may result in more money going toward temporary lodging, providing more public safety and human services, ultimately decreasing their revenue and increasing their expenditures. Investing in strategic disaster plans and teaching residents about disaster awareness, such as evacuation techniques and materials to purchase for homes and businesses, would help local governments save resources and become more financially resilient.

In addition, local governments must invest in proper infrastructure and enforce stricter building standards on developers. Stark (2002) argues that stricter building standards implemented in the wake of Hurricane Andrew may reduce future property damage by up to one-third. This is important because local governments rely on property taxes from residential and commercial property as a revenue stream to help pay for public services. Revenue received from property taxes may be limited in some cases, so local governments may also rely on debt mechanisms to assist with paying for public services. Investment in proper infrastructure is vital because a great deal of resources could be saved if local governments invest in disaster-resilient materials and technologies to reduce a disaster's impact. This will help to reduce the costs of providing shelter for displaced households, and it may potentially reduce the number of lives lost in the long run.

Last, local governments can financially prepare for natural disasters by pursuing business development that is resilient to the shocks of natural disasters. After a disaster, some businesses close temporarily, whereas some close permanently due to extensive costs of repair or rebuilding. Employment sectors such as construction and transportation are more resilient and tend to improve economic activity in the long run.

Conclusion

Despite the unpredictability of the damage caused by natural disaster, indicators of financial resiliency provide a framework for developing financial strategies that better prepare local governments for the consequences of such events. Local governments in disaster-prone areas would be well advised to develop policies devoted to promoting their financial resiliency. The chapter broadens our understanding of disasters' risk to the long-term economic wellbeing of communities and the need for financial preparation to mitigate the cost of such events.

Several important research questions merit further investigation by public-administration scholars. The most basic question is whether pre-impact spending on disaster preparation accelerates the return to economic normalcy following a disaster. For example, what effect do annual expenditures for emergency management have on the speed of recovery? Does the institutional location of the emergency-management agency in state government and the degree of centralization of emergency-management duties at the state versus local levels affect the return to normalcy?

Similarly, we lack empirical clarity on the effect of post-disaster spending on response and recovery on accelerating a community's return to normalcy. What effect does the amount of state and/or federal aid committed to responding to the disaster have on the return to normalcy? Does a local government's prior experience with similar disasters affect its post-disaster recovery? Are pre-disaster trends in changes in population and personal income exacerbated by a disaster?

These research questions raise fundamental issues about the administration of disasters and the role of institutional, budgetary and management practices in one of government's most critical roles—that of restoring economic and social normalcy in the wake of a disaster.

Bibliography

Alpert, B. (2006). Federal bill strips FEMA of housing duties; "trailer cities" also would be avoided. *New Orleans Times-Picayune*, 18 May, 16A.

Benson, C., and Clay, E. J. (2004). *Understanding the economic and financial impacts of natural disasters*. Washington, D.C.: World Bank Publications.

Bonanno, G. A. (2004). Loss, trauma, and human resilience: have we underestimated the human capacity to thrive after extremely aversive events? *American Psychologist, 59*, 20–28.

Bonanno, G. A., Rennicke, C., and Dekel, S. (2005). Self-enhancement among high-exposure survivors of the September 11th terrorist attack: resilience or social maladjustment? *Journal of Personality and Social Psychology, 88*, 984–998.

Campbell-Sills, L., and Stein, M. B. (2007). Psychometric analysis and refinement of the Connor-Davidson Resilience Scale (CD-RISC): Validation of a 10 item measure of resilience. *Journal of Traumatic Stress, 20*(6), 1019–1028.

S. Chandra, J. Acosta, S. Stern, L. Uscher-Pines, M.V. Williams, D. Yeung, et al. (2011). Building community resilience to disasters. RAND Corporation, Santa Monica, CA. www.rand.org/pubs/technical_reports/TR915.html. Accessed August 11, 2016.

Chang, S. E. (2003). Evaluating disaster mitigations: methodology for urban infrastructure systems. *Natural Hazards Review, 4*(4), 186–196.

Christopherson, S., Michie, J., and Tyler, P. (2010). Regional resilience: theoretical and empirical perspectives. *Cambridge Journal of Regions, Economy and Society, 3*, 3–10.

City of Coppell, Texas. (2014). Home Rule Charter for the City of Coppell. www.coppelltx.gov/government/city-secretary/city-charter. Accessed August 17, 2016.

Cutter, S. L., Boruff, B. J., and Shirley, W. L. (2003). Social vulnerability to environmental hazards. *Social Science Quarterly, 84*(2), 242–261.

Ewing, B. T., and Kruse, J. B. (2002). The impact of Project Impact on the Wilmington, North Carolina, labor market. *Public Finance Review, 30*(4), 296–309.

Ewing, B. T., Kruse, J. B., and Thompson, M. A. (2003). A comparison of employment growth and stability before and after the Fort Worth tornado. *Global Environmental Change Part B: Environmental Hazards, 5*(3), 83–91.

Fayol, H. (1918/1999). *Administration Industrielle et Générale*. Paris: Dunod.

Fergus, S., and Zimmerman, M. A. (2005). Adolescent resilience: a frame-work for understanding healthy development in the face of risk. *Annual Review of Public Health, 26*, 399–419.

Fowles, J., Liu, G., and Mamaril, C. B. (2009). Accounting for natural disasters: the impact of earthquake risk on California municipal bond pricing. *Public Budgeting & Finance, 29*(1), 68-83.

Gillespie, W. (1991). Economic impact of hurricane Hugo. *Division of Research and Statistical Services, Office of Economic Research*. Columbia: South Carolina Budget and Control Board.

Guimaraes, P., Hefner, F. L., and Woodward, D. P. (1993). Wealth and income effects of natural disasters: an econometric analysis of Hurricane Hugo. *The Review of Regional Studies, 23*(2), 97–114.

Gulati, A. G. (2013). Financial resilience as an integral part of disaster management: an essential for protecting the poor and vulnerable. *Kurukshetra, A Journal of Rural Development*, 61(12). Special issue, October 2013. Available at SSRN: http://ssrn.com/abstract=2337414. Accessed August 17, 2016.

Holling, C. S. (1973). Resilience and stability of ecological systems. *Annual Review of Ecology and Systematics*, 4, 1–23.

Jacobson, C., Hughey, K. F. D., Allen, W. J., Rixecker, S., and Carter, R. W. (2009). Toward more reflexive use of adaptive management. *Society & Natural Resources*, 22(5), 484–495.

Kendra, J., and Wachtendorf, T. (2003). Elements of resilience after the World Trade Center disaster: reconstituting New York City's Emergency Operations Centre. *Disasters*, 27, 37–53.

Killian, L. M. (1954). Some accomplishments and some needs in disaster study. *Journal of Social Issues*, 10(3), 66–72.

Kloha, P., Weissert, C. S., and Kleine, R. (2005). Developing and testing a composite model to predict local fiscal distress. *Public Administration Review*, 65(3), 313–323.

Kunreuther, H. (1996). Mitigating disaster losses through insurance. *Journal of risk and Uncertainty*, 12(2–3), 171–187.

LaLone, M. B. (2012). Neighbors helping neighbors: an examination of the social capital mobilization process for community resilience to environmental disasters. *Journal of Applied Social Science*, 6(2), 209–237.

Lemieux, F. (2014). The impact of a natural disaster on altruistic behaviour and crime. *Disasters*, 38(3), 483–499.

McGuire, M., and Schneck, D. (2010). What if Hurricane Katrina hit in 2020? The need for strategic management of disasters. *Public Administration Review*, 70(s1), s201–s207.

Office of Financial Management (February 11, 2014). Harris County Investment Policy p. 1–21.

Pittman, Elaine. (August 30, 2011). FEMA under fire as natural disasters pile up. *Emergency Management Magazine*. www.emergencymgmt.com/disaster/FEMA-Under-Fire-as-Natural-Disasters-Pile-Up.html

Pike, A., Dawley, S., and Tomaney, J. (2010). Resilience, adaptation and adaptability, *Cambridge Journal of Regions, Economy and Society*, 3(1), 59–70.

Pfeiffer, J., and Salancik, G. R. (1978). *The external control of organizations: a resource dependence perspective.* New York: Harper & Row.

Skidmore, M., and Toya, H. (2002). Do natural disasters promote long run growth? *Economic Inquiry*, 40(4), 664–687.

Stark, J. (2002). New building code brings cost, confusion. *St Petersburg Times*, 19. www.sptimes.com/2002/webspecials02/andrew/day2/story1.shtml. Accessed August 11, 2016.

Toya, H., and Skidmore, M. (2007). Economic development and the impacts of natural disasters. *Economics Letters*, 94(1), 20–25.

Walker, B., Holling, C. S., Carpenter, S. R., and Kinzig, A. (2004). Resilience, adaptability and transformability in social–ecological systems. *Ecology and Society*, 9(2), 5.

Wicks, A. G., Berman, S. L., and Jones. T, M. 1999. The structure of optimal trust: moral and strategic implications. *Academy of Management Review*, 24, 99–116.

Wildasin, D. E. (2009). *Intergovernmental Transfers to Local Governments.* Boston, MA: The Lincoln Institute of Land Policy.

Zedlewski, S. R. (2006). Pre-Katrina New Orleans: the backdrop. In Turner, M. A. and Ledewski, S. R. (eds.), *Rebuilding Opportunity and Equity into the New New Orleans After Katrina: The Urban Institute.* pp. 1–7. Washington, D.C.: The Urban Institute.

10 The Effects of Natural Disasters on Local-Government Finance

Orkhan Ismayilov and Simon A. Andrew

Introduction

Despite the important relationship between natural disasters and economic growth, few studies have explored the impact that natural disasters have on local-government finances (Pelling *et al.* 2002). The more quickly industries are able to rebuild and provide employment following a major disaster, the more quickly a local government will recover economically. In fact, a disaster may stimulate economic growth through the infusion of capital from disaster relief provided by state and federal agencies and casualty-insurance companies. This chapter uses a case study to compare trends in local sales-tax revenues in two Texas cities to assess the effect of disasters on local budgets.

Studying the effect of disasters on economic growth is critical in order to highlight how jurisdictions respond differently to support their economies. Examples of steps that are taken to bolster post-disaster economies are plentiful, with the rebuilding of old infrastructure among the most common. Often, local governments use this opportunity to reconsider their established policies and practices in order to recover and promote growth in a timely manner (Kousky 2009; Cuaresma *et al.* 2008; Pelling and Uitto 2001). Toya and Skidmore (2007) point out that Alaska and California have utilized sales taxes in order to increase revenue after natural disasters.

Baade *et al.* (2007) found that cities used tax revenues from industries during the reconstruction phase following a major disaster to reinvest in their infrastructure. After Hurricane Andrew, the City of Miami used net gain from wealthier parts of the city for rebuilding affected areas (Baade *et al.* 2007). In New Orleans, the financial assistance provided by the federal government and nonprofits was used by the city to rebuild the higher-priority areas that attracted tourists. Investment of resources to attract tourists stabilized and sustained economic growth because the city heavily relies on tourist-driven sales and gross receipt revenue. These actions highlight the importance of changing local-government economic policies and adopting new practices in order to raise revenue.

Other scholars argue that major disasters have a negative consequence on economic growth through loss of productive capital (Baade *et al.* 2007; Carter *et al.* 2007) and that the natural disasters' consequences for business activities vary depending on the length of the recovery processes as well as the type of disaster (Otero and Marti 1995; Toya and Skidmore 2007). The literature also suggests that certain natural disasters do not have a significant effect on economic activities; rather, the assistance provided by governmental and non-governmental organizations is the key element to economic recovery (Strömberg 2007).

Natural disasters can be devastating for a local-government economy, and assistance provided by the federal government is not usually sufficient to fully recover. While governments seek to rebuild damaged properties and provide services, they are also faced with the reality of increased maintenance costs and a decision to "build back better" than before the disaster. The reality that local governments compete with each other to attract citizens and businesses is made stark as this competition intensifies following a major disaster, when households are often displaced. During this time local governments scramble to retain their workforce, major industries and commercial centers. To further examine these events, this chapter studies the behavior of two local governments in Texas with respect to the economic strategies they adopted in rebuilding following major disasters. We examine both cities' patterns of sales-tax revenue between 2000 and 2014 as a way to gauge how their economies responded to disasters.

The goal of this study is to identify relationships between disaster and economic growth by investigating the techniques that local governments use to promote growth. To accomplish this, we first examine the literature related to natural disasters and economic growth. Next we outline the methods employed in choosing the two cities that are the subject of this study and discuss the major threads related to their sales-tax revenues. Finally, we discuss the strategies adopted by the City of Beaumont in stimulating the city's business activities and ways it has coped with natural disasters. The final section concludes the chapter and examines this chapters limitations and suggests future research areas.

Disasters, Economic Growth and Recovery

Natural disasters have a negative effect on economic growth (Strömberg 2007; Raddatz 2009). The extent of negative consequences vary depending on the type and scale of disaster and are associated with loss of capital stock, highlighting the importance of retaining industrial and commercial capital for economic growth after disasters. Other elements, such as debt and income levels, also play major roles in determining how a jurisdiction will fare post-disaster (Melecky and Raddatz 2011; Kellenber and

Mobarak 2008; Guimaraes *et al.* 1993). A similar observation was made by Noy (2009), who highlighted the fact that workforce retention in the aftermath of a storm is critical. Without sufficient personnel and resources to repair damages and build new infrastructure, a region is likely to experience a slow economic recovery. One method of speeding up this recovery will undoubtedly be increased revenue.

Other scholars have written on the potential positive effects of disaster on the local economy. Based on the Gravity model of trade, Cuaresma *et al.* (2008) demonstrated through a cross-country analysis of technology transfer that an increase of trade and resources exchanged is often spurred after disasters. For example, when a government does not have the resources to invest in new technology, it will increase its level of technological trade with other countries.

A study of the relationship between economic development and major disasters (Guimaraes *et al.* 1992) used ARIMA modeling and time-trend analysis to investigate the effect of hurricanes on wealth and income. This model assumes that natural disasters generally help an economy in the short run because of a surge in construction and retail activities. The findings of this study suggest that construction projects provide local residents with employment opportunities, while retail activities simulate the local economy through purchasing of materials and supplies to repair structural damages.

Disasters and Local-Government Finance

The consequences of disasters can be observed through the patterns of local taxes and revenues. Picou and Martin (2006) found that, after Hurricane Ivan in 2004, damaged cities located along the Gulf Shores and Orange Beach, raised taxes to provide incentive for investment in the local economy and to make it more productive for the long term. Studies also found that relying on taxes was one of the key influences that helped Alaska and California recover from natural disasters (Toya and Skidmore 2007). Increased tax revenue boosts local-government growth by creating a multiplier effect, whereby an increase in revenue creates growth in other areas of local government. Another reason for short-term economic recovery is the recovery assistance and financial resources provided by state and federal governments, as well as aid received from emerging relief efforts. Local governments must be able to effectively manage all of the aid provided from the various sources (Skidmore and Toya 2002).

In addition, local governments also adapt to retain and attract new business. One argument is that local governments adopt policies and practices to raise revenue and continue providing public services. According to Crater (2014), in Houston, the assistance given to local governments after Hurricane Ike accounted for $1–10 million for larger businesses and anywhere from $10 to $150,000 for smaller businesses between 2011 and 2012. Crater (2014) also notes that it is important for small businesses to get assistance and reopen their businesses, because in the long run, local governments rely on local businesses for tax revenue. Assistance provided to start private businesses has a positive effect on economic growth and revenue that local governments use during the recovery phase.

While an increase in sales and gross sales might have a short-term effect on local-government finance, in the long run it is unclear whether businesses from non-affected areas locate branches in affected areas. Companies often shy away from high-risk areas due to fear of losing capital investment and disruption of supplies and production. While disasters increase sales and employment in affected areas in the short term, it is unclear whether the initial public assistance provided by the federal and state governments will have positive results in the long term (Raschky 2008). Studies have shown that states that have a stronger economy prior to the natural disaster tend to recover faster in the short term than states with a weaker economy (Loayza et al. 2009). Similarly, a disaster-affected region that is not financially distressed before a disaster occurs is likely to benefit economically during the recovery phase (Cavallo et al., 2010). Farmers who make their living off the land are most affected, since they are no longer able to grow their agricultural products; however, workers or non-agricultural laborers who make their living through reconstruction and rebuilding generally will benefit financially from a disaster (Cavallo et al. 2010).

Whereas research on disaster mitigation, planning, preparedness and recovery has been conducted, few studies have investigated the effect of disasters on local-government economy, economic growth and development. One argument is that immediate assistance provided by the federal and state governments to victims and disaster-affected local governments provides a sufficient amount of resources for economic recovery and growth. However, Chang (1984) argues that whereas assistance provided by the federal government is crucial for affected cities to recover after disasters, local governments are not able to maintain the level of assistance provided by federal and state governments over the long term.

Additionally, scholars argue that greater openness and competition should be promoted among local governments (Toya and Skidmore 2007). Application of rainy-day funds and effective programs helps local governments to grow by applying the Harrod–Domar growth model, whereby a stronger economy that has been impacted by disaster attracts investment and promotes growth. That is, if a government has a strong economic base, including a surplus and adequate disaster funding, then private institutions invest more aggressively in these areas because they can expect a return on their investments. For example, if a local government has had previous disasters but managed to recover and grow, then private institutions view this behavior as investment-worthy. Toya and Skidmore (2007) provide evidence for this argument by showing economic growth in small governments with greater openness and competition.

Governments that face civil and financial difficulty experience negative consequences and longer-term economic recovery after natural disasters (Benson and Clay 2004). Moreover, a government whose economy is solely based on a single industry will find economic recovery difficult post-disaster. Therefore, types of disasters and types of local-government economies go hand in hand. For example, a local-government economy that is reliant on the retail industry is less likely to be affected by wildfires, while a local-government economy that is reliant on the farming industry may be more likely to be affected.

Chang (1984) reiterates that hurricanes can have a devastating effect on development and growth. For instance, Hurricane Frederic in 1979 caused devastating damage to the southern region of Alabama. Overall damage caused by the disaster, without adjustment for inflation, amounted to $2.3 billion dollars. The federal government allocated aid to counties, which helped increase short-term growth. According to Chang (1984), sales-tax revenue increased in the coastal region of Alabama, which is explained by an increase in the retail trade and construction sectors. In addition, the retail trade and construction sectors increased temporary employment. At the same time, local governments adopted business-friendly policies to attract private businesses. While Chang (1984) does not fully discuss the effects of business-friendly policies on the local-government economy, he highlights that economic growth was short-term. Hence, local governments experienced quick increases in sales tax and income, but failed to retain these gains in the long term.

Furthermore, local governments should plan for disasters by area, anticipating their economic effects. French *et al.* (2010) highlight that disaster-prone local governments should plan for decreased property-tax revenue after natural disasters. According to Benson and Clay (2004), governments need appropriate risk-management strategies and financial planning for an 8–10-year timeframe. Since the risk of devastating disasters is low for many local governments, many managers and administrators hesitate to spend funds and conduct adequate financial preparation for disaster, as funds are always needed elsewhere.

Background: How Do Disasters Affect Local-Government Finance?

The frequency of natural disasters in Texas is among the highest in the U.S. (FEMA 2013). Since 1954, 3,562 Presidential disaster declarations have been issued for natural disasters. Between 2000 and 2013, the number of total natural disaster declarations was 2,202 (FEMA 2013). Notably, eighty-three events have been major-disaster declarations (FEMA 2013). At $16.9 billion, the cost of natural-hazards insurance payouts in Texas between the years 2000 and 2013 was greater than that of any other state (Christian 2014). With 72,131 extreme events declared between 1950 and 2010, the frequency of natural hazards is greater than that of all other types of disaster, and not every disaster is declared an emergency.

According to Musgrave and Musgrave (1973), local governments' sales taxes make up a large proportion of local-government revenue in the U.S. On average, local governments' sales and gross receipt taxes account for 34 percent of revenue (Tax Foundation 2013). In order to help residents in affected areas rebuild damaged infrastructure and property, state and local governments make certain products and services exempt from sales tax. It is assumed that sales-tax exemptions can negatively affect local governments' sales-tax revenue because this signifies a decrease in revenue.

Sales tax is also directly related to business activities during the reconstruction and recovery phase of disasters because, as previously noted, purchases and charges for repairing damaged property and rebuilding infrastructure are tax-exempt (Texas Comptroller of Public Accounts 2016). Moreover, the Texas Comptroller of Public Accounts (2016) indicates that labor charges associated with damage repair are also exempt from sales tax.

According to the U.S. Census Bureau Tax Foundation (2013), local and state government tax revenues consist of property tax (35 percent), sales and gross receipt tax (34 percent), individual income tax (20 percent), corporate tax (3 percent), motor vehicle license tax (2 percent) and other taxes (6 percent). However, the state of Texas does not impose income tax, which means local governments rely heavily on property and sales tax. Two main sources of income for local governments, especially cities, are property and sales tax. Property tax is the largest source of revenue for local governments, whereby local governments determine a value for property and a rate that is applicable for one year. Hence, property tax is collected on a yearly basis. However, the rate does not change during the year; even if a natural disaster happens mid-year, the rate and revenue will not change until the year ends.

Sales tax, on the other hand, is imposed on retail sales, leases and rentals, whereby revenue changes daily. This means sales-tax revenue is income-elastic: if income increases then sales increase as well, which also leads to an increase in sales-tax revenue. As a broad-base tax, sales tax relies on sales of goods and services that fluctuate daily, weekly and monthly. Hence, sales-tax revenue can increase and decrease at a faster phase compared to property-tax revenue. The authors determine that sales tax becomes a crucial source of revenue for local governments because the increase in sales-tax revenue occurs faster than that in property-tax revenue. Moreover, due to population migration and disaster damage in affected areas, local governments decrease the property-tax rate to retain their population. According to French *et al.* (2010), property-tax revenue decreases after natural disasters, whereas Chang (1984) finds that sales-tax revenue increases after natural disasters. Therefore, the authors will investigate sales-tax revenue because, in order to increase revenue after disasters, local governments have leverage to raise their sales tax revenue.

Case Studies: Beaumont and Waco

The City of Beaumont is located in Jefferson County, along the coast line of the Gulf of Mexico. Since 2000, the county has had six major-disaster declarations. The devastation brought about by Hurricane Ike in 2008 was estimated to result in about $28 billion of economic damages (Texas Engineering Extension Service/Storm Resource 2010). According to the Texas Engineering Extension Service (2010), twelve months after Hurricane Ike, the county suffered economic losses of around $11.9 billion. The manufacturing sector was most affected, comprising $9.1 billion of the loss. The disaster also affected the agricultural sector, with nearly 4,000 head of cattle reported drowned during the Sabine River basin storm surge flood. Debris removal for the county was estimated to cost around $36 million. The amount of public and individual assistance provided to Beaumont since the year 2000 is estimated at about $72 million (FEMA 2013). Much of the damages have been paid by the private sector, including insurance payouts (Texas Engineering Extension Service/Storm Resource 2010).

The City of Waco's experience is somewhat different. The City of Waco is located in McLennan County. Since 2000, the county has had three major disaster declarations by the federal government. In 2013, the county experienced a fertilizer explosion, resulting in fifteen deaths, more than 150 injured and damage to more than 140 properties (Merchant and Mone 2013). Employment and business activities were lost in he area surrounding the disaster. However, the City of Waco itself has not experienced any major disasters that have affected the local economy even though there were three major disaster declarations recorded—(hurricane, fire and flood). With the exception of a few tornadoes that caused damage of up to a few million dollars, the city has not experienced any catastrophic disasters. However, before 2000, the city experienced fire suppression and emergency-management declarations. It is important to note that although events are declared as disasters, not all major-disaster declarations are for catastrophic disasters. The authors have focused on choosing cities that have or have not experienced catastrophic natural disasters.

Table 10.1 shows the comparable demographic characteristics of both cities, that is, population density, education, age and race. Population density, percentage of older people and land area are very similar for both cities. We also notice that both cities have a similar percentage of college-educated population and average number of people per household. Two differences in population were noted including the median age of citizens in Beaumont being higher than that of Waco as well as a difference in racial composition of the two cities. For instance, Beaumont's population consists of almost equal percentages of Black and White people, and a small percentage of Hispanic people. Waco's population consists of majority White people, with an equal percentage of Black and Hispanic people. To summarize, the authors find that cities have similar demographic characteristics with the exception of their racial composition.

Additionally, Table 10.1 shows economic characteristics such as income, housing, employment and sources of revenue for both cities. Income characteristics such as median family income and median nonfamily income are similar for both cities. Median household income, per capita income and percentage below poverty level are relatively similar. It is important to note that income characteristics are slightly higher for Beaumont than for Waco. In addition, while both cities have the same sales-tax rate, sales tax as a source of revenue is twice as high for Beaumont than for Waco. Table 10.1 shows that Beaumont heavily relies on sales tax and property tax as revenue sources, while Waco relies on utility and property taxes as sources of revenue.

Housing and employment characteristics are similar for both cities. While the homeownership rate is slightly higher in Beaumont, we note that housing units, medial house value and number of households are similar for both cities. Underemployment characteristics, employed population and unemployed population are similar for both cities. The number of firms and retail sales per capita is slightly higher in Beaumont.

The pattern of sales-tax revenue in Beaumont and Waco can be examined by the timeline of natural disasters between 1990 and 2013 and the patterns of sales-tax revenues (see Figures 10.1 and 10.2). According to FEMA (2013), in the third quarter of 2005 and the third quarter of 2008, Jefferson County was declared a major-disaster area. Hurricanes Katrina[1] and Rita of 2005 and Hurricane Ike of 2008 caused damage to thousands of properties in the City of Beaumont and the federal government assisted the affected areas by providing $72,372,207 million in federal assistance. The sales-tax revenue per number of business establishments also shows an external shock during these two periods.

Based on the patterns of sales-tax output, we note that the City of Beaumont experienced irregular revenue changes when compared to the City of Waco. In Figure 10.1, we see that the total amount of sales tax per business establishment in Beaumont increased in the third quarters of 2005 and 2008, but the pattern shows an increase of 54 percent in the third

Table 10.1 Demographic and Economic Characteristics: Beaumont and Waco

	Beaumont	Waco
Population Density		
Population	117,796	129,030
Population Density (per sq. mile)	1,428	1,403
Percentage Population 65 and Older	12%	11%
Land Area in Square Miles	83	89
Water Area in Square Miles	3	12
Education and Age		
Bachelor Degree and Higher	23%	21%
Persons per Household	2.5	2.6
Median Age	35	28
Income		
Median Household Income	$40,765	$32,239
Mean Household Income	$60,769	$46,840
Median Family Income	$49,550	$40,376
Median Nonfamily Income	$27,189	$21,052
Per Capita Income	$24,310	$17,846
Percent Below Poverty Level	22%	30%
Housing		
Housing Units	50,689	51,452
Percentage Homeownership Rate	58%	46%
Median House Value	$98,100	$91,800
Households	45,512	45,087
Employment		
Population Employed	50,334	52,571
Population Unemployed	6,063	5,074
Private Business Workers	38,730	43,453
Government Workers	8,879	6,475
Total Number of Firms	9,943	7,912
Retail Sales Per Capita	$22,568	$16,068
Race		
Percentage Black	47%	21%
Percentage White	40%	51%
Percentage Hispanic	13%	29%
Revenue		
Sales Tax Rate	8.25%	8.25%
Sales Tax	32%	13%
Property Tax	29%	24%
Utility Revenue	9%	38%
Industrial and Business Payments	12%	4%

Sources: U.S. Census 2010, 2012; City of Beaumont Operating Budget 2012; City of Waco Operating Budget 2012

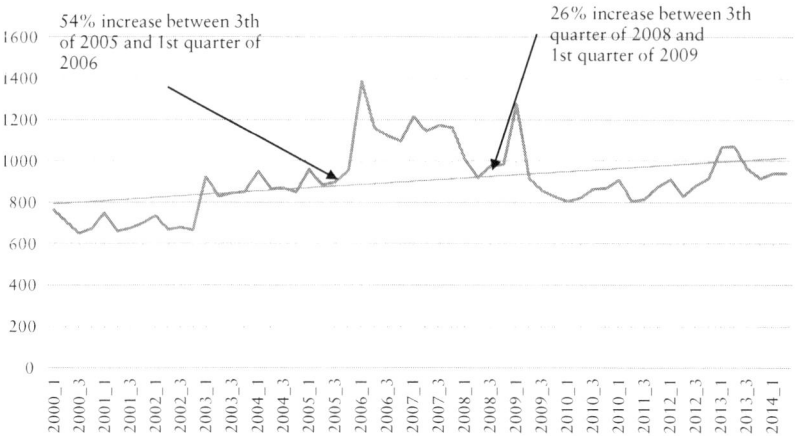

Figure 10.1 Sales Taxes Per Number of Business Establishments—Beaumont, TX
Sources: US Census estimates 2000–12, Texas Comptroller's Office Public Records

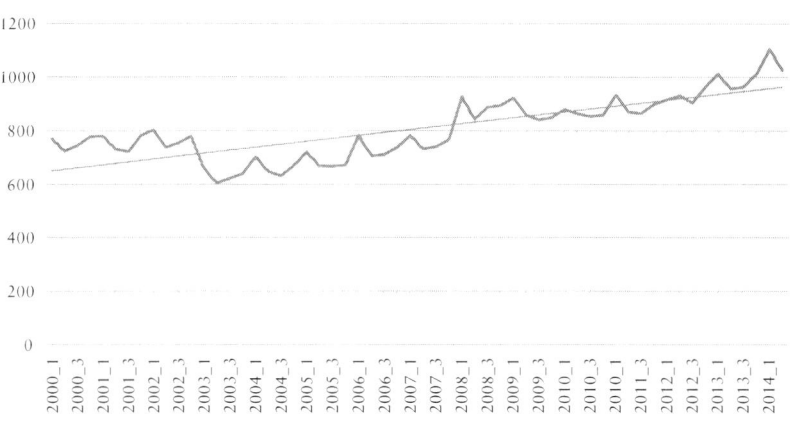

Figure 10.2 Sales Taxes Per Number of Business Establishments—Waco, TX
Sources: US Census estimates 2000–12, Texas Comptroller's Office Public Records

quarter of 2005 as well as during the second quarter of 2006. A similar pattern occurs between the third quarter of 2008 and the first quarter of 2009, with, an increase of 26 percent in sales-tax revenue. However, in the City of Waco within the same period of time, the patterns of sales-tax output for each quarter shows gradual annual revenue growth of 16

The Effects of Natural Disasters 219

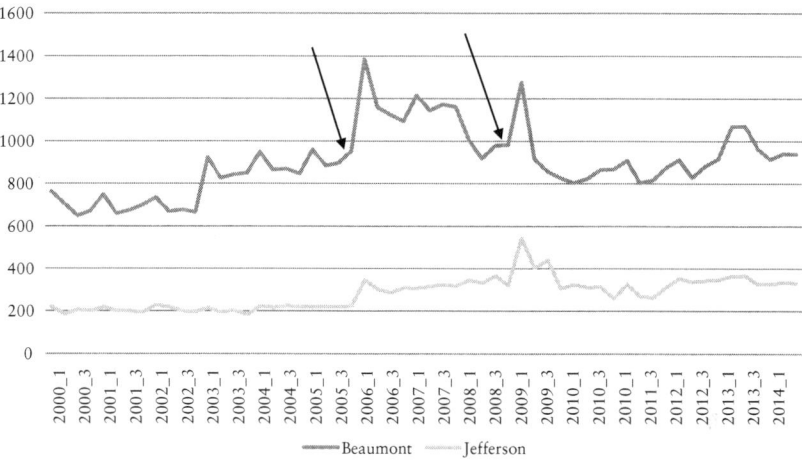

Figure 10.3 Sales Taxes Per Number of Business Establishments—Beaumont (city) and Jefferson (county), TX

Sources: US Census estimates 2000–12, Texas Comptroller's Office Public Records

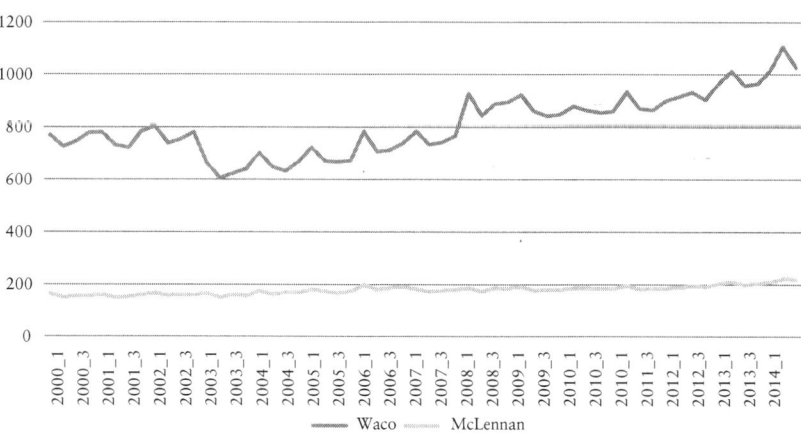

Figure 10.4 Sales Taxes Per Number of Business Establishments—Waco (city) and McLennan (county), TX

Sources: US Census estimates 2000–12, Texas Comptroller's Office Public Records

percent. The pattern of the total amount of sales-tax revenue shows a slight decline in the fourth quarter of 2002, but then slowly recovers. It is important to note that no major disaster happened in Waco during this period and the amount of sales tax per number of business establishments has been growing at the expected level of 8–14 percent for the past fourteen years.

There are two possible explanations as to why sales-tax revenue increases during the short term after an external shock. First, the assistance provided by the federal government provides temporary relief, allowing residents in the affected area to purchase necessary goods to repair damage caused by disasters. According to FEMA (2013), the initial amount of assistance provided by the federal government as temporary relief to the disaster affected region was over $70 million. The state government sales-tax exemption for certain products and services also enables residents to purchase goods (Picou and Martin 2006).

Insurance claims for Jefferson County between 2000 and 2008 accounted for over $390 million (TWIA 2014), while flood insurance claims for Beaumont accounted for over $40 million (FEMA 2014), allowing local residents to purchase products to repair damage and contributing to the rise in revenue in Beaumont. In addition, both public and private sectors provided business loans to small business to assist in maintaining employment and commerce. According to the Texas Engineering Extension Service/Storm Resource (2010), the amount of loans provided to Jefferson County amounted to more than $39 million dollars. Housing and individual grants for qualified residents and private businesses in the affected area accounted for over $50 million dollars (Texas Engineering Extension Service/Storm Resource 2010).

Local policymakers in the Beaumont–Port Arthur metropolitan area provide economic incentives for industries and businesses, which attract new investment and create new jobs. Because the petrochemical industries provide more than 40 percent of county paychecks, the city officials adopted incentives (mainly tax incentives) to promote growth. For instance, by providing economic incentives such as state grants (e.g. $2.1 million from the Texas Enterprise Fund) and local incentives, the City of Beaumont has been able to attract industries such as Orascom Construction Industries, adding more than 3,000 construction jobs and 240 permanent jobs in the city (Shannon 2013; Wallach 2013).

Following Hurricanes Rita and Katrina, the City of Beaumont experienced a 1 percent growth in the overall non-farm employment category between 2005 and 2008, which represents an average of 1,475 workers per year (US HUD 2008). The unemployment rate for Beaumont ending in May 2008 averaged 5.5 percent annually, slightly higher than the national average of 5.4 percent (US HUD 2008). The city lost about 8,000 non-farm jobs six months after Hurricane Rita due to a decline in production in the petrochemical manufacturing industry. However, between 2005 and 2008 the city's non-farm goods employment category increased by 2.4 percent, mostly led by hiring in mining, construction and natural resources. The non-farm employment category also saw an increase of 4,100 jobs, representing over 70 percent of overall job gains in the City of Beaumont (US HUD 2008).

Beaumont's housing and rental markets have experienced negative consequences after natural disasters. For instance, the greater Beaumont area had a vacancy rate of 1.5 percent post-Hurricane Rita and approximately 10,650 single-family homes and 8,900 mobile units were destroyed.

In the three years after Hurricane Rita, however, housing prices increased by 3 percent due to the permanent loss of 480 apartment units and damaged to 4,400 additional units. During the 2006 to 2008 time period the rental vacancy rate decreased by 4 percent and the demand for multifamily permits quickly accelerated, leading to an increase in single-family house prices.

Sales-tax revenue saw a decline in the third quarter of 2009. The financial impact of disaster can be felt if the sales-tax revenue is significantly affected by a loss of population, especially the loss of skilled workforce (Curtis *et al.*, forthcoming). This pattern is common in disaster affected regions. In New Orleans after Hurricane Katrina, for example, over half of the population migrated elsewhere and did not return (Koerber 2006). Tax-credit incentives used by New Orleans attracted industries and increased employment, thus affecting the long-term growth of tax revenue (Lopez 2012; Liu and Plyer 2007). According to the Department of Housing and Urban Development (2008), migration of residents and households in counties along the Gulf Coast increased in the mid-2000s due to Hurricane Katrina (Curtis *et al.*, 2013). In Beaumont, between 2005 and 2006, the population decreased by 1.7 percent due to Hurricane Rita (US HUD 2011). The pattern was similar in 2007, with a population decrease of 0.2 percent. Moreover, the Texas Comptroller of Public Accounts (2016) indicates that labor charges associated with damage repair were also exempt from sales tax.

The Greater Beaumont Chamber of Commerce (GBCOC) is a regional chamber that has allocated economic support to over 1,000 businesses following natural disasters. A grant of up to $1 million provided by the Texas government was managed under the South Texas Economic Development Foundation (STEDF), with up to $15,000 in zero-interest loans available to eligible businesses. According to the GBCOC, following Hurricanes Rita and Katrina in 2005, $945,325 was awarded in small-business loans. A total of $758,600 was awarded to small businesses by the Texas Disaster Relief Fund, with a return of $503,924. Another $186,725 was awarded by the South East Texas Disaster Relief Fund, with a return of $145,350.

CDCs are nonprofit organizations, certified under the Small Business Administration 504 Program (SBA), which provide assistance to small businesses by financing communities through expansion of the tax base, business growth and job creation. Loans are provided to businesses to purchase or renovate land, buildings and equipment. The STEDF is a certified economic development organization which is an affiliate of the GBCOC, composed of the Texas counties of Jefferson, Orange and Hardin. According to the GBCOC, $100,000 was given to the STEDF by the South East Texas Emergency Relief Fund (SETERF) to provide assistance to social-service agencies and help fund loans to small businesses following Hurricane Rita.

Following Hurricane Ike, Southeast Texas was given $190 million by the Texas state government. According to the GBCOC, the amount of $1.9 million was provided to small businesses in the form of forgivable loans, with the condition that they stayed open for at least two years after the disbursement. In addition, rural areas received assistance from Texas government. Following Hurricane Ike, 769 applications totaling $11,936,500 were made by small businesses in rural areas. In total, 108 applicants were successful, with an outlay of $1,710,000 in forgivable loans. The conditions applied to small businesses play a big role in small-business development and local-government finance through new businesses' contribution to the tax pool.

In order to gain information from the disaster-affected City of Beaumont, the authors contacted city officials such as the Director of Economic Development and the President of the Chamber of Commerce. Moreover, the authors requested city-council meeting minutes between the years 2003 and 2013 and requested reports on economic development. While the information disclosed was not detailed, the authors were able to find sufficient information on the types of policies and strategies used by the affected city administration. The authors were unable to investigate the economic policies and strategies adopted by the City of Waco's administration because the city's source of revenue—sales tax—never changed. The growing pattern of sales-tax revenue for the City of Waco is slow but steady. Hence, due to the lack of a significant pattern in sales-tax revenue for the City of Waco, the authors decided not to further investigate the economic policies and strategies adopted by the city administration.

Lessons learned

In the City of Beaumont, after natural disasters—specifically hurricanes—we noticed an increase in sales-tax revenue from the third quarter of 2005 to the second quarter of 2006. A similar pattern occurred between the third quarter of 2008 and the second quarter of 2009. The case study provides findings that are supportive of the literature on short-term economic growth during the post-natural disaster period. However, increased sales-tax revenue could be directly related to the assistance and loans provided by the private and public sectors to qualifying residents and small businesses, implying that initial monetary backing increases sales-tax revenue for local governments.

Strömberg (2007) and Crater (2014) note that assistance provided by the public and private sectors is crucial for the economic development of affected areas. The authors indicate that this assistance, in the form of loans and grants, is an essential tool during the recovery stage. While assistance is provided to affected local governments by the federal and state governments, regional chambers of commerce, economic development organizations and certified development companies play a major role in allocating assistance to small businesses. Assistance provided to small businesses in the form of forgivable loans and block grants is allocated by regional organizations.

Skidmore and Toya (2002) also note that governments can raise revenue by investing in human capital. The authors suggest that policymakers create a worker-friendly environment through rebranding in order to increase human capital, which enables affected areas to increase capital. Policymakers could encourage industry to relocate to the affected area by relaxing local rules. Beaumont policymakers, in collaboration with their county and state counterparts, were able to offer tax exemptions for certain goods, products and services. Furthermore, when initial temporary exemptions expired, policymakers were able to prolong exemptions, which helped the city economy in the short run. In addition, policymakers offered grants and tax incentives to businesses and industries which in turn added new jobs and promoted economic growth for the long term.

Aside from receiving aid from state and federal governments, local governments rely on aid from the private sector, such as insurance companies, highlighting the importance of private-sector participation in economic development. FEMA notes that, as of 2014, City of Beaumont residents have been paid almost $50 million in flood insurance alone. According to Melecky and Raddatz (2011), governments that were economically developed prior to a natural disaster generally experienced an expansion of revenue deficit. However, insurance claims prevent large revenue deficits by providing resources to disaster-affected areas. Hence, insurance benefits play a major role in economic development after a natural disaster. This argument aligns with our findings in Beaumont. Insurance claims helped the City of Beaumont to recover after a natural disaster by rebuilding damaged property, preventing permanent migration and increasing sales in the city, highlighting the significant role of the private sector in local governments' economic development and growth.

Previous findings indicate that administrators and managers must adequately prepare themselves for disasters by preparing budgets allowing for the possibility of decreased revenue of one type with the possibility of increasing another type revenue (Benson and Clay 2004). For instance, French et al. (2010) highlights that natural disasters have a negative effect on property-tax revenue in the long run, meaning governments must prepare themselves for a long-term decline in revenue. In addition, Toya and Skidmore (2007) indicate that governments' openness and disclosure of information on disaster-preparedness promotes growth because businesses and industries invest in areas where the risks are low.

Traditionally researchers argued that assistance and aid promotes only short-term growth led by stabilizing growth in an overall economy (Lindell and Prater 2003). According to Chang (1984), local governments experience increased sales-tax revenue shortly after disasters due to an inflow of insurance and governmental assistance. Furthermore, Chang (1984) notes that local governments are not able to retain most of the revenue generated by spending occurring in affected areas. This case study highlights that there is a surge in sales-tax revenue shortly after disaster occurs, but the sales-tax revenue decreases over time.

This case study not only supports the findings of previous scholars but also provides an overview of the policies and strategies adopted by an affected city to increase revenue. Sales-tax revenue, as mentioned earlier, is a revenue on retail sales that fluctuates daily. That is, if a local government sees an increase in business, then sales-tax revenue increases as well. Since sales-tax revenue makes up over 32 percent of the City of Beaumont's revenue, city administrators are interested in promoting business activity in order to increase revenue. This case study highlights that a business-friendly environment can be created through public and private partnerships.

Public and private partnerships can be achieved in various ways; the City of Beaumont's case demonstrates that one effective way is to provide assistance through tax abatements and loans. An important strategy adopted by the City of Beaumont was neighborhood empowerment zones, created by providing incentives to businesses and residents such as building-fee waivers, expedited permit reviews, lien waivers and construction tax abatements. The purpose of tax abatements is to improve residents' quality of life, improve the local economy by enhancing the value of the local tax base and increase economic activity. As a result, empowerment zones attracted businesses and promoted growth.

To summarize: in our case study the affected city, which is heavily dependent on sales-tax revenue, adopted policies and incentives to promote economic growth. Local governments that are prone to natural disaster should prepare themselves economically for disasters. Depending on the severity of the disaster, administrators should consider adopting policies and strategies in the immediate aftermath. Although during the recovery stage governmental assistance and insurance claims increase local-government finances, the main task is retaining money devoted to affected areas. The authors find that local governments that promote economic growth, through tax abatements and public–private partnerships, are able to increase sales-tax revenue.

Conclusion

This study investigated the effect of natural disasters on local-government finance. The goal was to find the relationship between disaster and economic growth by investigating techniques that local governments use to promote growth. We investigated the patterns of sales tax, that is, the difference in sales-tax revenues before and after disasters. We also researched the economic strategies and practices which Beaumont adopted to recover and promote economic activity after Hurricanes Rita, Katrina and Ike. It is important to keep in mind that this study investigates sales-tax revenue after major disasters declared by the federal government.

The City of Beaumont experienced a surge in sales-tax revenue after major natural disasters that occurred in 2005 and 2008. The authors did not investigate the effect of types of disasters on local-government economies; however, the two natural disasters experienced by Beaumont were in fact catastrophic disasters. According to the Saffir–Simpson Scale, 1 being minimal and 5 being catastrophic, Hurricane Ike was rated level 4 and Hurricane Rita was rated level 5, showing that both natural disasters were rated highly on the scale of wind damage. In addition, Hurricane Ike accounted for $28 billion in damage and Hurricane Rita accounted for over $11 billion in damage incurred.

While the authors do not compare the effects of natural disasters based on a scale of disaster effects, we highlight that only two major disasters occurred in the affected City of Beaumont and both were catastrophic. Hurricane Rita (2005) had a greater economic impact on Beaumont than Hurricane Ike. Figure 10.1 showed that the sales-tax revenue surge after Hurricane Rita was higher and longer-lasting than that in the aftermath of Hurricane Ike. Nonetheless, both disasters were catastrophic and accounted for billions of dollars in damage to the affected city. Since no other non-catastrophic disasters occurred in Beaumont between 2000 and 2006, the authors cannot make a prediction on the effect of disasters based on a scale of the effects and differences in economic policies adopted by the local government.

The findings of this study suggest that through city and regional governments' provision of economic incentives such as grants, and local tax incentives, Beaumont was able to attract businesses and industries to increase sales-tax revenue. The literature indicates that disasters have both a positive and a negative effect on an economy. However, the effects of natural disasters are determined based on the type of a disaster. Hochrainer (2009) notes that in the short term, climatic disasters generally have a negative effect on economic development. Contrary to what Hochrainer found, this study shows sales-tax revenue increases within a few months of a natural disaster.

However, we found that the pattern of sales-tax revenue in the City of

Beaumont after natural disasters fell behind the estimated sales-tax revenue growth. The findings highlight that major disasters could have an effect on sales-tax revenue in the long run because, according to the pattern, revenue decreases. Our findings also suggest that decreased sales-tax revenue in the long run could be related to relocation of businesses and population, because relocation of industries affects employment. Moreover, the information provided by city officials suggests that Beaumont offered tax exemptions to businesses and industries soon after their disasters in order to retain business activity and attract industries from other areas with a goal of increasing local-government revenue and local employment opportunity.

There are a few limitations to this study. Although it highlights that Beaumont used tax exemptions and information-openness tactics to attract businesses, we are uncertain regarding the other strategies Beaumont used to gain private-sector support. For future research, it will be necessary to contact the governments of both Beaumont and Waco to gather information on strategies used for this purpose after disasters. It is likely that by contacting both cities, we will be able to locate details that are not included in city reports and public records.

Future studies should also explore different types of disasters and their impact on tax revenues. That is, the effect of natural disasters on local-government finance varies by type. For instance, studies conducted by Raddatz (2009) and Hochrainer (2009) investigating disasters' impact on economic development found that different types of disasters have a different effect on economic growth. According to Raddatz (2009), climatic disasters such as droughts, hurricanes and coastal storms reduce the economy by 1 percent or more. Hochrainer (2009) notes that aid received does not have any significant effect on economic development of affected areas.

In conclusion, this study highlights the need to study sales-tax revenue after natural disasters and the practices adopted by local governments to increase sales-tax revenue. The study suggests that sales-tax revenue increases when assistance, in the form of block grants and forgivable loans, is provided by the public sector. We also find, however, that most of the assistance comes from the private sector, in the form of insurance claims. By getting assistance from the public and private sectors, and by adopting strategies such as providing economic incentives in the form of grants and tax exemptions, local governments are able to increase sales-tax revenue and recover after natural disasters.

Note

1 The City of Beaumont was not directly affected by Hurricane Katrina but saw an increase in evacuees from New Orleans.

Bibliography

Baade, Robert A., Robert Baumann, and Victor Matheson. "Estimating the economic impact of natural and social disasters, with an application to Hurricane Katrina." *Urban Studies* 44, no. 11 (2007): 2061–2076.

Benson, Charlotte, and Edward J. Clay. *Understanding the economic and financial impacts of natural disasters*. Washington, D.C.: World Bank Publications, 2004.

Carter, Michael R., Peter D. Little, Tewodaj Mogues, and Workneh Negatu. "Poverty traps and natural disasters in Ethiopia and Honduras." *World Development* 35, no. 5 (2007): 835–856.

Cavallo, Eduardo, Andrew Powell, and Oscar Becerra. "Estimating the direct economic damages of the earthquake in Haiti." *The Economic Journal* 120, no. 546 (2010): F298–F312.

Chang, Semoon. "Do disaster areas benefit from disasters?" *Growth and Change* 15, no. 4 (1984): 24–31.

Christian, Carol. "Texas and Oklahoma lead nation in costliest natural disasters" (2014). www.chron.com/news/houston-texas/texas/article/Texas-and-Oklahoma-lead-nation-in-costliest-5252032.php. Accessed August 5, 2016.

Crater, Dana. *Case studies in small business finance following a disaster*. International Economic Development Council (2014). http://restoreyoureconomy.org/case-studies-small-business-finance-following-disaster. Accessed August 5, 2016.

Crespo Cuaresma, Jesús, Jaroslava Hlouskova, and Michael Obersteiner. "Natural disasters as creative destruction? Evidence from developing countries." *Economic Inquiry* 46, no. 2 (2008): 214–226.

Curtis, Katherine, Elizabeth Fussell, and Jack DeWaard. "Do migration systems predict post-disaster migration patterns? The case of the Gulf of Mexico coastal counties before and after Hurricanes Katrina and Rita." Paper presented at the Population Association of America 2013 Annual Meeting. http://paa2013.princeton.edu/abstracts/130346. Accessed August 5, 2016.

Federal Emergency Management Agency. "Disaster declarations for 2013" (2013). www.fema.gov/disasters/grid/year/2013. Accessed August 22, 2016.

French, Steven P., Dalbyul Lee, and Kristofor Anderson. "Estimating the social and economic consequences of natural hazards: fiscal impact example." *Natural Hazards Review* 11, no. 2 (2010): 49–57.

Guimaraes, Paulo, Frank L. Hefner, and Douglas P. Woodward. "Wealth and income effects of natural disasters: an econometric analysis of Hurricane Hugo." *The Review of Regional Studies* 23, no. 2 (1993): 97–114.

Hochrainer, Stefan. "Assessing the macroeconomic impacts of natural disasters: are there any?" *World Bank Policy Research Working Paper Series, Vol* (2009).

Kellenberg, Derek K., and Ahmed Mushfiq Mobarak. "Does rising income increase or decrease damage risk from natural disasters?" *Journal of Urban Economics* 63, no. 3 (2008): 788–802.

Koerber, Kin. "Migration patterns and mover characteristics from the 2005 ACS Gulf coast area special products" (2006). In *Southern Demographic Association Conference, Durham, North Carolina*.

Kousky, Carolyn, Olga Rostapshova, Michael Toman, and Richard J. Zeckhauser. "Responding to threats of climate change mega-catastrophes." *World Bank Policy Research Working Paper Series, Vol* (2009).

Lindell, Michael K., and Carla S. Prater. "Assessing community impacts of natural disasters." *Natural Hazards Review* 4, no. 4 (2003): 176–185.

Liu, Amy, and Alisson Plyer. "A review of key indicators of recovery two years after Katrina: second anniversary special edition" (2007). www.brookings.edu/research/reports/2007/08/katrinarecovery-liu. Accessed August 5, 2016.

Loayza, N., E. Olaberria, J. Rigolini, and L. Christiansen. "Natural disasters and medium-term economic growth: the contrasting effects of different events on disaggregated output." *World Bank Policy Research Working Paper Series*, Vol (2009).

Lopez, Adrianna. "A look into America's fastest growing city" (2012). www.forbes.com/sites/adrianalopez/2012/07/26/a-look-into-americas-fastest-growing-city/. Accessed August 18, 2016.

Liu, Amy, and Allison Plyer. *A review of key indicators of recovery two years after Katrina* (Washington, DC: Brookings Institution, 2007).

Melecky, Martin, and Claudio Raddatz. "How do governments respond after catastrophes? Natural-disaster shocks and the fiscal stance." *Natural-Disaster Shocks and the Fiscal Stance (February 1, 2011), World Bank Policy Research Working Paper Series*, Vol (2011).

Merchant, Nomaan, and John Mone. "Fertilizer plant explosion in Texas levels building, claims as many as 15 lives" (2013). https://web.archive.org/web/20130419070218/www.huffingtonpost.com/2013/04/18/fertilizer-plant-explosion-texas_n_3106023.html. Accessed August 5, 2016.

Musgrave, Richard, and Peggy Musgrave. *Theory of Public Finance* (New York: McGraw-Hill, 1973).

Noy, Ilan. "The macroeconomic consequences of disasters." *Journal of Development Economics* 88, no. 2 (2009): 221–231.

Otero, Romulo Caballeros, and Ricardo Zapata Marti. "The impacts of natural disasters on developing economies: implications for the international development and disaster community." In *Disaster Prevention for Sustainable Development: Economic and Policy Issues*, ed. Mohan Munasinghe and Caroline Clarke (Washington DC: World Bank, 1995): 11–40.

Pelling, Mark, Alpaslan Özerdem, and Sultan Barakat. "The macro-economic impact of disasters." *Progress in Development Studies* 2, no. 4 (2002): 283–305.

Pelling, Mark, and Juha I. Uitto. "Small island developing states: natural disaster vulnerability and global change." *Global Environmental Change Part B: Environmental Hazards* 3, no. 2 (2001): 49–62.

Picou, J. Steven, and G. Cecelia Martin. *Community impacts of Hurricane Ivan: a case study of Orange Beach, Alabama*. Boulder, CO: Natural Hazards Center (2006).

Raddatz, Claudio. "The wrath of God: macroeconomic costs of natural disasters." *World Bank Policy Research Working Paper Series*, Vol (2009).

Raschky, Paul A. "Institutions and the losses from natural disasters." *Natural Hazards and Earth System Science* 8, no. 4 (2008): 627–634.

Shannon, James. "Largest US methanol plant to be built in Beaumont". *The Examiner* (2013). http://theexaminer.com/stories/news/largest-us-methanol-plant-be-built-beaumont. Accessed August 5, 2016.

Skidmore, Mark, and Hideki Toya. "Do natural disasters promote long run growth?" *Economic Inquiry* 40, no. 4 (2002): 664–687.

Strömberg, David. "Natural disasters, economic development, and humanitarian aid." *The Journal of Economic Perspectives* 21, no. 3 (2007): 199–222.

Texas Comptroller of Public Affairs. "Local sales and use tax" (n.d.) www.window.state.tx.us/taxinfo/local/index.html. Accessed August 5, 2016.

Texas Engineering Extension Service/Storm Resource (Knowledge Engineering) "Hurricane Ike Impact Report; Executive Summary" (2010). www.thestormresource.com/Resources/Documents/county_reports/Executive_Summary.pdf. Accessed August 5, 2016.

Texas Windstorm Insurance Association. "Hurricane Damage Facts" (2014). www.twia.org/. Accessed August 5, 2016.

Toya, Hideki, and Mark Skidmore. "Economic development and the impacts of natural disasters." *Economics Letters* 94, no. 1 (2007): 20–25.

Toya, Hideki, and Mark Skidmore. "Natural disaster impacts and fiscal decentralization." *CESifo Forum* 11, no. 2 (2010), pp. 43–55.

United States (U.S.) Census Bureau. "State & County Quickfacts" (2013). http://quickfacts.census.gov/qfd/index.html. Accessed August 5, 2016.

United States Department of Housing and Urban Development (US HUD). Office of Policy Development and Research "Housing Recovery on the Gulf Coast: Phase II" (2011). www.huduser.gov/publications/pdf/gulfcoast_phase2.pdf. Accessed August 5, 2016.

Wallach, Dan. "Updated: New OCI plant means 3,000 construction jobs" (2013). www.beaumontenterprise.com/news/article/Updated-New-OCI-plant-means-3-000-construction-4998313.php. Accessed August 5, 2016.

Index

absenteeism 153
acquisition regulations 17
Act for the Relief of Sick and Disabled Seamen (1798) 61
action plans 76, 151–3
active mitigation 116
ADA National Network 64, 75
adaption 197
adaptive measures 116
AECOM 28–9
Affordable Care Act (2010) 63
AFN (access and functional needs): assistance for 75; categories of 60–1; definition 64–5; emergency planning 60; focus on most vulnerable groups 66; immobility 66; planning, resources required for 65–6; transportation needs during Hurricane Katrina 68, 69; *see also* people with disabilities; special needs
Agency Chief Human Capital Officers 136
aid flows, federal 36–7
AIR Worldwide 176, 177
Alesch *et al.* 110–11
alternative risk transfer products 171–2
American Federation of the Physically Handicapped (1940) 63
American National Standards Institute (1958) 63
American Red Cross 181
Americans with Disabilities Act (1990) 63
Americans with Disabilities Act (2008) 64
Americans with Disabilities Act (2009) 64, 67
Architectural Barriers Act (1968) 63
ARIMA modeling and time-trend analysis 211
Army Corps of Engineers (ACE) 39, 96

Baade *et al.* 209
Baja California peninsula 179
Barbour Administration 41
Base Closing Commissions 52
Beaumont case study: demographic characteristics 215–16; devastation brought by Hurricane Ike 215; economic characteristics 216; employment/unemployment 216, 220–1; flood insurance claims 220, 223; Greater Beaumont Chamber of Commerce (GBCOC) 221; housing market 216, 221; loans to Jefferson County 220; neighbourhood empowerment zones 224; population decrease 221; sales tax revenue 216, 218–20, 221, 222, 225–6; tax exemptions 223, 224
Benson, C. and Clay, E.J. 202, 213
Biggert–Waters Flood Insurance Reform Act (2012) 2, 49; revised 2013 181
Birkland, T.A. 67
Birkland, T.A. and DeYoung, S.E. 4, 6
Birkland, T.A. and Nath, R. 120
Blaikie *et al.* 118
Blanck, P. 62
Blanco, Governor 68, 69
Bloomberg administration 48
Bonanno, G.A. 196

Index 231

bonding social capital 93, 94
bonds: General Obligation 202–3; municipal 203; ratings 203; Revenue 203
Boston Consulting Group 48
Boswick, Lieutenant General 50
bottom up recovery 93–6
bridging social capital 93, 94
Bring New Orleans Back Commission 95, 97
Brinkley, D. 72
Broadmoor Improvement Association 95
Budgetary Control Act (BCA) (2011) 189
budget-stabilization funds 203
Build it Back program 48
Bureau of Labor Statistics 102
Bush, President G.H.W. 55n3
business activity, categorizing by type of 111
businesses: impact of Hurricane Sandy 21–4; interaction with government 17; needs during disasters 19–21; operations during crises 17; and public procurement 16, 17; *see also* small businesses

Cadwalader, Wickersham & Taft 176
Calderone, T. 96
Calhoun County, Alabama (case study) 127–30; disaster-recovery phase 129; donated labour and resources 129; FEMA guidelines and federal disaster procurement regulations 127; human-capital component of disaster recovery 130; impacted by tornado 127; local government administration 127; long-term recovery phase 129; securing an outside consultant 127; volunteer-registration centers 128
Calhoun County Commission 127
California 138–9, 160–1
California War Council 138
Cantrell, L. 95
Card, A. 55n3
Carter, President, J. 182
catastrophe reinsurance 187
catastrophic natural disasters 215
cat bonds (catastrophe bonds) 171–2; case study *see* MultiCat Mexico 2009 Catastrophe Bond; CatMex 176; contingency funding to the DRF 189–90; high-quality securities 174; Interest Spread 177; Original Principal Amount 177; purchased from foreign governments by the U.S. Treasury 190–1
CatMex cat bond 176
CDCs (Certified Development Companies) 221
Cellco Partnership (Verizon) 28
Central Cocos 176
centralization of emergency management 1–4; alterations to the intentions of federalism 2–3; disaster declaration process 6–7; event-driven centralization 6; history of relief spending 4–5; overview of methods 4–7; using policy as incentive 3–4
Chamlee-Wright, E. and Storr, V.H. 94, 95, 96, 97
Chang, S.E. 201, 212, 213, 224
Chang, S.E. and Falit-Baiamonte, A. 111, 118
Chevron U.S.A. 42
Cicek, B. 96
citizens, role during a disaster 131
City Assisted Evacuation Plan 71
City of Los Angeles (Disability Rights Legal Center 2011) lawsuit 73
City of Miami 209
City of New Orleans 36
City of Oakland, CA (Disability Rights Advocates 2010) lawsuit 73
Civil Rights Act (1964) 67
Civil Service Reform Act (CSRA) 133
Cleveland, President G. 181
Clinton, President, B. 186
Coast Guard 26, 39
Coggburn, J.D. 134
collateral trust funds 174
collective-action problem 87, 88, 90, 97
collective bargaining 137, 159
communication: departmental 154–5; with outside stakeholders 155; special needs registries 77
Community Development Block Grants (CDBG) 44–5
Community Emergency Response Teams (CERT) 78

community leaders: Doris Voitier 95–6; Father Vien 94–5; initiators of local recovery 88, 89, 94–6, 98; LaToya Cantrell 95
community rebound: bridging social capital 94; importance of local leaders 89; incorporating lessons from 96–7; initiatives led by community leaders and entrepreneurs 98; rebuilding of communities 97; study of 88–9
community resilience 197
community risk profile 121
community volunteer groups 129
competition: fair 17–18; full and open 16–19, 24–5, 26–8, 31; public procurement 16, 17
comprehensive emergency management 112
Congress, OIC proposals 51–2
Consolidated Appropriations Act (2014) 50
contingency funds 203
contingent financing 175
contractors: inflated bids of 43; and public procurement 16
contracts: adjustments to process of 20; favouring established businesses 28, 30; FPDS-NG report 25–9; full and open competition 16–19, 24–5, 26–8, 31; non-competitive 18; post-disaster, bidding for 15; transparency 16, 32
coping actions 116
Corey, C.M. and Deitch, E.A. 110
County Business Patterns 21, 24
Crabill, A. and Rademacher, Y. 93, 98
Crater, D. 211–12, 223
crises: business operations during 17; flexible and adaptable approach to 135, 136, 137; hiring policies during times of 153; IT issues during times of 153; local governments' response to 36; personnel reform as a result of 135–8
crisis-management theories 120
Cuaresma et al. 211
Cutter et al. 118, 201
czars 52–3

Dahlhamer, J.M. and D'Souza, M.J. 110
Dahlhamer, J.M. and Tierney, K.J. 113
Danes et al. 110, 111
Dauber, M.L. 5
Davis-Bacon Act 44
debris removal functions 142
debt: local 202–3; long-term 202–3
Declaration of Independence 61
declarations for disaster 6–7
Department of Homeland Security (DHS) 91; Agency Chief Human Capital Officers 136; collective bargaining 137; creation of 135, 159–60; criticisms of 160; discretion in managing human capital 160; full discretionary powers of director of 136; grant programs 4; Office of State and Local Government Coordination and Preparedness 182; reform movement 137
Department of the Army 26
Department of Transportation Disadvantaged Business Enterprises program 106
Dietch, E.A. and Corey, C.M. 111
disability: definition 62, 64; discrimination 62; interpretation of term 64; specific conditions meeting criteria of 64; *see also* people with disabilities
Disabled Union Army 62
Disaster Acquisition Response Team (DART) 20
disaster-contingency fund, Texas 203
disaster literature, small businesses 102
disaster management: effective 90; failure of government post-Katrina 90–1; *see also* post-disasters
Disaster Mitigation Act (2000) 38
Disaster Mitigation Fund (DMF) 191
disaster recovery: AECOM missions 29; disaster relief 39; effective government planning 91; importance of community leaders to 94–6; specialist knowledge 137; spending 50; *see also* Calhoun County, Alabama (case study); MultiCat Mexico 2009 Catastrophe Bond
Disaster Relief Act (1974) 38, 182

Disaster Relief Appropriations Act (2013) 39
Disaster Relief Fund (DRF) 183–6, 189–90; funding and budgeting practices 184–6
Disaster Research Center (DRC) 109; studies 118
disaster risk financing 180–3; Disaster Relief Fund (DRF) 183–6, 189–90; disaster relief to foreign governments 190–1; domestic disaster relief funds 189–90; funding for domestic disaster mitigation programs 191; Hazard Mitigation Grant Program (HMGP) 183; National Pre-Disaster Mitigation Fund (PDMF) 183; new methods 187–8; parametric triggers 190; proactive, ex ante plan 188–91; structure 188–9; USAID's Office of Foreign Disaster Assistance (OFDA) 186–7
disaster service workers 138, 149
discrimination 62, 67
District of Columbia (Disability Rights Advocates 2014) lawsuit 73
domestic emergencies, management of 6
Donoghue, A.K. and Joyce, P.G. 4
Drabek, T.E. 117, 118, 120
Dust Bowl 139
Dynes, R.R. 131

earthquakes: Baja California peninsula 179; financial risk 203; Haiti 87; Northridge, California 54; San Francisco (1906) 181
Economically Disadvantaged Women Owned Small Businesses (EDWOSB) 108
economic growth: natural disasters and 209, 210, 210–11; recovery and 210–11
economic resilience 197
education: emergency planning for people with disabilities 78; of employees 155–6
Education for All Handicapped Children Act (1975) 63
Edwards, F.L. and Afawubo, I. 131
EIS (Environmental Impact Statement) 45

emergency management: budget cuts 65; business mitigation activities, research on 115–16; business responses, research on 116–17; business sector 118, 119; business size 117–18; categorizing small business in research 111; centralization of see centralization of emergency management; current issues with special needs registries 74–6; education 78; implications for policy on small businesses 120–1; implications for practice of 121–2; legal perspectives 66–7; limited resources for mitigation or preparedness projects 65; local agencies 121; preparedness of businesses, research on 113–15; recovery of businesses from disaster, research on 112–13; short-term and long-term setbacks 197; small businesses, research on 109–19; special needs registries 72–4; see also disaster management; post-disasters
emergency managers: acknowledging the importance of human capital 142; analyzing workforce requirements 148, 149, 150, 151; collaboration for preparedness 154; developing action plans for the HR function 151, 152; disaster service workers and 138–9; education and training for emergency planning 155–6; identifying human capital gaps 155; immobility, consideration of 66; lack of support from local politicians 157–8; managing human capital 154, 155; maximizing compatibility and shared goal-setting 141; obstacles faced by 156; planning meetings 145, 147; special needs registries, reluctance to use 75; working with businesses 121; working with HR professionals 156; working with others 144; working with the private sector 147
Emergency Operations Center (EOC) 69
emergency operations plans (EOPs) 140

emergency planning: by department 148; departmental communication 154–5; education on 155–6; HCM best practices in local government 142–3, 151; integrating human capital management and local government 139–42; management of volunteers 150–1; meeting locations for support personnel 150; meetings 145–6, 147, 148, 151, 155; needs and capabilities of all stakeholders 147–8; networking 147; non-essential employees 150; obstacles 156; outside groups 146–7; private sector as partner 147; skills inventories 150; tabletop exercises 147; teams 143–4, 145, 161; training of employees 149–50, 152, 155–6, 157; understanding human capital management 148–9; working with others 144–5

Emergency Planning and Community Right to Know Act (1986) 146

emergency preparedness: actions 116; working with others 144–5

emergency-response phase 128

emergency support functions (ESFs) 140

employees: counselling during states of emergency 152–3; demands on post-Katrina 131; education for emergency planning 155–6; essential 140, 149, 150, 159; group decision-making 143–4; impact on trust of 134; individual rights and collective bargaining 133; non-essential 138, 140, 149, 150, 162; salaries 131; training for emergency operations 149–50, 152, 155–6, 157; understanding the capabilities of 148–51; *see also* human capital; personnel

employment: job market shifts 199–200; unemployment 200

Entergy 95

entrepreneurs, initiators of local recovery 88, 89, 94, 98

equilibriums 196, 197

Esnard, A-M. and Sapat, A.K. 19

essential employees 140, 149, 150, 159

Evacuating Populations with Special Needs (Federal Highway Administration) 75

evacuation: inadequate planning and enforcement during Hurricane Katrina 47; late 68; orders 47; planning 71–2

Executive Order 13632 47

farmers 212

Fayol, H. 196

Federal Acquisition System (FAR) 16, 17, 29; exclusions 18; fair competition 17–18; full and open competition 17–18; interaction between business and government 17; rules allowing consideration for previous experience 29; sole sourcing 18

federal disaster-management policy: delegating local initiative and authority 35–6; protecting democratic-republican norms and values 35; revision of 35; *see also* mega-disaster

Federal Emergency Management Agency 26

federal funds/grants 1; awarding authorities' efficiency targeting specific firms 30; Community Development Block Grants (CDBG) 44–5; dependency on 3, 4; as incentive 3–4; and national goals 3; politicised process 7; pre-authorized disaster-relief 39; as solution to social problems 5; *see also* FEMA (Federal Emergency Management Agency)

federal government: assistance to AFN individuals 65; centralization of emergency management *see* centralization of emergency management; collective-action problem 90, 97; delegation of authority and decision-making to local authorities 97–8; disaster expenditure 170; flexible and adaptable approach to crises 135, 136, 137; human capital as an asset 137; institutional memory as an asset 137; interaction with business 17; intertwining of disaster-relief funding and general welfare spending 5; the knowledge problem 90; outcry against, post-Hurricane Katrina 6; public procurement *see* public procurement; realistic and achievable outcomes 97–8; recognition of community involvement in local recovery 89; response to Hurricane Katrina 39; stress caused by disasters 159; transparent decision-making 19; using grants for implementing national goals 3; various sources and management agencies of aid 36–7

Federal Highway Administration 46, 75; evacuation plans for people with disabilities 76; funds 43–4

Federal Insurance & Mitigation Administration (FIMA) 191

federalism: alterations to the intentions of 2–3; defense of 35

Federal Labor Relations Authority 137

Federal Procurement Data System—Next Generation (FPDS-NG) 24–9

Federal Railroad Administration 46

Federal Transit Administration Emergency Relief Program 39, 46

FEMA (Federal Emergency Management Agency): alleged poor performance of 160; authority and responsibilities of 38; contract oversight during major disasters 20; creation of 182; definition of AFN 64–5; disaster mitigation funding 191; disaster-preparedness and recovery efforts 92; Disaster Relief Fund (DRF) 183–6, 189–90; Federal Insurance & Mitigation Administration (FIMA) 191; funding for domestic mitigation efforts 191; funds diverted for Hurricane Irene 197–8; Hazard Mitigation Grant Program (HMGP) 183; hurricane management 6; loss of responsibility for disaster mitigation grants 182–3; mimicking best practices 96; misallocation of labour and resources 91; mitigation planning 38; National Pre-Disaster Mitigation Fund (PDMF) 183; obtaining funding for disaster relief 188; partnerships with state and local stakeholders 92; partnership with AECOM 29; perceived powers of 6; PKEMRA guidelines 91–2; Public Assistance (PA) funds 43–4; Public Assistance Technical Assistance Contracts 29; recognition of mistakes 92; reduced capacity to act 1; replacement vs. improvement conflicts 43; restructuring of 38–9; strategic human capital management 136–7; temporary housing for Hurricane Katrina victims 39–40; trailers for the community 95; Whole Community concept 2, 89, 92

Ferleger, D. 62

financial costs, natural disasters: earthquakes in Mexico 173; of global warming 170; of hurricanes 170; insured losses 169; natural disasters in America 170–1; of tornadoes 170; uninsured losses 169; *see also* cat bonds (catastrophe bonds); disaster risk financing; insurance; reinsurance

financial, insurance and real estate (FIRE) sector 118, 119

financial management, local government 129–30

financial resiliency: concept of 197–8; indicators of *see* indicators of financial resiliency; of local governments 198; of state governments 198

Flood Disaster Protection Act (1973) 181, 182

flooding 181

flood insurance 181; rates 2; subsidies 49

Florida Hurricane Catastrophe Fund (FHCF) 187

Florida Special Needs Working Group 72–3

Florida Statutes § 252.355, Title XVII 72

focusing events 67

Fonden 173–4, 175, 177; Trust 174

Fopreden (Natural Disaster Prevention Fund) 173–4

forecasting 152

236 Index

forgivable loans 222, 223
Fothergill, A. and Peek, L.A. 66
Fowles *et al.* 203
Fox *et al.* 66
Franklin, B. 61
French *et al.* 213, 223
French, P.E. and Goodman, D. 134
Fugate, C. 89, 91–2
fund balance 203

Gabris *et al.* 134
GAO (Government Accountability Office) 2; contracting review 20; institutional memory 135–6
gender, small businesses and 108
General Obligation bonds 202–3
general welfare spending 5
Gillespie, W. 199
Glenn, D. 96
global urbanization 169
global warming 170
global wealth, increasing 169
Goldman Sachs 176
Gordon, P. and Ikeda, S. 97
GO Zone 41; beneficiaries of 42; bonds and tax credits 42–3; economic plan favouring large investors 42–3; evaluation of investments 42; first-come, first-served approach 42
grant funds *see* federal funds/grants
grants administration, local government 129, 130
Gravity model of trade 211
Greater Beaumont Chamber of Commerce (GBCOC) 221, 222
Great Lakes Dredge and Dock Company 29
group decision-making 143–4
Gulati, A.G. 197
Gulf Coast Recovery Project 88–9

Haiti, earthquakes 87
Harrod–Domar growth model 212
Hayek, F.A. 90
Haynes *et al.* 110, 111
hazard adjustments 116
hazard assessment 116
hazard-education programs 121
hazard mitigation 116
Hazard Mitigation Grant Program (HMGP) 183, 191

Hewitt, P. 74, 75
high reliability theory 120
Historically Underutilized Business Zones (HUBZone) 108
Hochrainer, S. 225, 226
Hoffman, S. 66
Homeland Security Act (HSA) (2002) 136
Homeowner Flood Insurance Affordability Act (2014) 50
Hopkins, S. 61
Houston *et al.* 78
Howe, P.D. 110
HR (human resources) 129; *see also* human resource management (HRM)
HR professionals 151–2, 154, 155, 156
human capital: availability from general public 131, 137–8; case study, Calhoun County 127–30; and the DHS 160; HR professionals' understanding of 154; implications of disaster on 131; increased amount and variety of 155; management and disasters, state and local governments experience with 138–9; managers' reliance on 135; preserving institutional memory 135–6, 137; private sector 138; recognition of value of 130; strategic management of 135; *see also* employees; human-resource management (HRM)
human capital management (HCM): analyzing workforce requirements to achieve strategic direction 148–51; best practices in local government emergency planning 142–3; building relationships 159; commitment to collaboration 155; departmental representation at meetings 158; environment of preparedness 158–9; identifying the strategic direction of the government unit 143–8; and local government emergency planning 139–42; strategic 161
human resource management (HRM): collective bargaining 159; developing action plans for the HR function 151–3; emphasis on merit

and procedures 133; environment of meritocracy 159; impact on employees 134–5; managers' reliance on human capital 135; National Performance Review (NPR) 133; New Public Management (NPM) 133–4; obstacles faced by personnel 156; personnel reform as a result of crisis 135–8; political patronage 133; in the public sector 132–5; public servants 132; spoils system 132–3; strategic leadership 134; transfer of responsibilities to first-line managers 133; *see also* employees; human capital; personnel
Hurlbert *et al.* 94
Hurricane Alex 179
Hurricane Andrew 54, 169, 171, 182, 187, 209
Hurricane Celia 202
Hurricane Frederic 213
Hurricane Gustav 70, 71, 72
Hurricane Hugo 169, 171, 199
Hurricane Ike 187, 211–12, 215; rating on the Saffir–Simpson Scale 225
Hurricane Ingrid 179
Hurricane Irene 197
Hurricane Ivan 211
Hurricane Katrina 6, 7, 39–46; Barbour Administration response 41; cleanup activities of Mississippi state 40–1; Coast Guard response 39; Community Development Block Grants (CDBG) 44–5; deaths 87; designated disaster zone 41, 42; diffuse program favouring large investors 41–2; evacuation planning post-hurricane 71–2; federal government failures 20, 41; federal government response to 39; FEMA response to 39–40; financial damage 54, 87; GO Zone *see* GO Zone; inadequate evacuation planning and enforcement 47; late evacuation 68; lawsuits arising from 73; local government response to 40; National Guard response to 40; planning for future hurricanes 70–1; policy authoritativeness, federal government failure 41; policy integration, federal government failure 41; prevention of recurrence 49; Public Assistance (PA) funds 43–4; replacement vs. improvement conflicts 43; requests for flexibility in funding expenditure 44; Road Home initiative 45–6; special needs issues 67–72
Hurricane Katrina: A Nation Still Unprepared (2006) 69
Hurricane Manuel 179–80
Hurricane Mitch 172
Hurricane Odile 180
Hurricane Pam study 67–8
Hurricane Rita 42, 136, 160, 221; rating on the Saffir–Simpson Scale 225
Hurricane Sandy 7, 21–4, 46–50; Build it Back program 48; bureaucratic delays 49; deaths and financial damage 87; evacuation orders 47; failure of federal government's long-term recovery efforts 48; financial cost 201; FPDS-NG report 24–9; leadership failures 48–9; Rapid Repairs program 48; rebuilding efforts, failure of 48; recovery program, failures of 48–9, 50; restoration of vital services 47
Hurricane Sandy Rebuilding Strategy 47
Hurricane Sandy Rebuilding Task Force 47
hurricanes, timeline 178
Hurricane Wilma 172
Hy, R.J. and Waugh, W.L. Jr. 147

ICS (Incident Command System) 152, 157
immigrant community 94
immobility, AFN group 66
indicators of financial resiliency 198–9; budget reserves and contingency funds 203; job market shifts 199–200; local debt 202–3; local revenues and expenditures 202; long-term investment in infrastructure 201; management 204; management use of 204–5;

market value 200; per capita income 201; population trends 200; property and casualty insurance 202 infrastructure 201; investment in 204, 205
institutional memory 135–6, 137
insurance: casualty 202; claims for flooding in Beaumont 220, 223; as a consequence of Hurricane Odile 180; flood 181; losses 169; natural-hazards payouts 214; property 202

Jackson, A. 132
Jacksonville State University 142
Jacobsen et al. 140, 197
Jindal, B. 72
John C. Stennis Institute of Government 142

Kailes, J.I. and Enders, A. 61, 64, 66
Kell, J. 40
Keyes, W.N. 18
Kloha et al. 199
knowledge problem, the 90
Koppelin, A. 44
KSAs (knowledge, skills and abilities) 152

lawsuits: City of Los Angeles (Disability Rights Legal Center 2011) 73; City of Oakland, CA (Disability Rights Advocates 2010) 73; District of Columbia (Disability Rights Advocates 2014) 73; New York City (Santora & Weiser 2013) 73; Washington, DC (Beck 2014) 73
leadership unity 141
League of Physically Handicapped (1935) 62
Leeson, P.T. and Sobel, R.S. 7
Lehne, R. 17
Levering Act (1950) 138
Lewis et al. 141
Liberty Zone 41
Lindell, M.K. 146
Lindell, M.K. and Perry, R.W 116
linking social capital 93–4
local emergency planning committees (LEPCs) 146, 161
local government finances 209–10; Beaumont case study see Beaumont case study; business-friendly policies 213; competition for citizens and businesses 210; disasters and 211–13; economic growth after natural disasters 209, 210, 210–11; effect of disasters on 213–15; financial assistance provided by federal government 212; Harrod–Domar growth model 212; openness and competition 212–13; planning by area 213; positive effects of disaster on local economies 211; recovery and 210–11; recovery linked to economic strength 212; reliance on specific industries 213; retaining and attracting new business 211–12; sales tax exemptions 214; tax revenues 209, 211, 214; Waco case study see Waco case study
local governments: altering modes of operation 4; analyzing workforce requirements to achieve strategic direction 148–51; backing up law-enforcement efforts 36; collaboration for further preparedness 154, 155; compliance with FEMA policy and guidelines 129–30; counselling of employees during state of emergency 152–3; demands on, post-Katrina 130–1; departmental representation at meetings 158; dependency on federal grants 3, 4; emergency-response phase 128; employees' lack of specialist knowledge 160; experience with human capital management and disasters 138–9; finance, effects of natural disasters on see local government finances; financial management 129–30; financial resiliency see financial resiliency; financial responsibility for disaster response and recovery 198; financial strain 131; grants administration 129, 130; HCM best practices in emergency planning 142–3, 151; hiring of temporary personnel 129; hiring policies during times of crisis 153; HR professionals 151–2, 154, 155, 156; human capital available from the general public 131; human-

capital gap 137; impacts on human resources 129; impaired decision-making due to central funding 3; increasing financial resiliency 198; integrating human capital management and emergency planning 139–42; IT issues during times of crisis 153; lack of support from local politicians 157–8; limited supply of employees 131; management of volunteers 151; outsourcing 129; priority activities in a disaster 127–8; property taxes 131, 204–5; recipients of vast government funding 3; recognising the value of human capital 130; responses to crises 36; setting up volunteer-registration centers 128; strategic direction during disaster response 143–8

local recovery 87–90; collective-action problem 87, 88, 90, 97; community rebound 88–9, 94; disaster assistance from top down 90–3; federal government intervention required 88; federal government recognition of community involvement 89; inadequate resources for 88; incorporating lessons from successful community rebound 96–7; initiatives of local entrepreneurs and community leaders 88, 89, 94–6, 98; policy implications 96–8; rebuilding of communities 97; recovery from the bottom up 93–6

Long Beach Earthquake 139
Longmore, P.K. 62–3
long-term recovery phase 129
Lott, Senator 41–2
Louisiana Department of Transportation and Development (DOTD) 68
Louisiana Hurricane Task Force 68–9
Louisiana National Guard 40
Louisiana Office of Homeland Security and Emergency Preparedness (LOHSEP) 68
Lovell, C.H. 3, 4
Lynn, D. and Klingner, D.E. 133

Maestri, W. 68

Mann, S. 142
Maryland, defining small businesses 104–6, 108
Maryland Minority Business Enterprise (MBE) Program 108
Mary Queen of Vietnam Catholic Church 89, 91, 94
mega-disaster: avoiding excessive centralization of authority 35; centralizing authority 36; government aid 37; local government initiatives 36; OIC (Officer in Charge) 37; response and recovery 36
mental illness 61–2
Mexico: disaster risk financing 173, 173–4; federal budget 173, 174; Ministry of Finance 173, 188; MultiCat Mexico 2009 Catastrophe Bond see MultiCat Mexico 2009 Catastrophe Bond; natural hazards 172–3; susceptibility to earthquakes 172–3
Mexico City 176
Ministry of Finance (Mexico) 173; obtaining funding for disaster relief 188
Minority Small Business and Capital Ownership Development Program ("8(a) program") 106–8
mitigation: active 116; defining 116; Disaster Mitigation Fund (DMF) 191; disaster mitigation funding 191; emergency management phase 38, 48, 50, 65, 115–16; Hazard Mitigation Grant Program (HMGP) 183, 191; National Pre-Disaster Mitigation Fund (PDMF) 183, 191; passive 116
Montes de Oca, P.H. 116
Montgomery County Local Small Business Reserve (LSBR) 105–6
Moody's 180
Moynihan, D.P. 135, 136
MultiCat Mexico 2009 Catastrophe Bond: earthquake parameters 175–6; issuance timeline 178–9; launching 176; Offering Circular Supplement 178; parametric triggers 175–6, 176, 177; performance of 179–80; structure of 177–8

MultiCat program 172; establishment of 176; expanding use of 190; generic 178; *see also* cat bonds, MultiCat Mexico 2009 Catastrophe Bond
Munich Re Capital Markets 176
municipal bonds 203
Musgrave, R. and Musgrave, P. 214

Naff *et al.* 137
Nagin, Mayor 68, 69, 71
National Association of State Procurement Officials 18
National Commission for Reconstruction 173
National Easter Seal Society (1958) 63
National Federation of the Blind (1940) 63
National Flood Insurance Plan (NFIP) 170, 181
National Guard 40
National Hurricane Center 54, 68
National Oceanic and Atmospheric Administration (NOAA) 200
National Park Service 26
National Performance Review (NPR) (1993) 133
National Pre-Disaster Mitigation Fund (PDMF) 183, 191
National Response Framework 140
National Response Plan (NRP) 70
National System of Civil Protection (SINAPROC) 173
natural disasters: effect on local government finance *see* local government finances; financial costs *see* financial costs, natural disasters
Natural Disasters Fund (1996) 173
Navy Seabees 39
neighbourhood empowerment zones 224
networking 147
New Jersey: contract value afforded vendors 26; housing recovery program 49; impact of Hurricane Sandy 21, 24
New Orleans: income from tourism 209; population migration 221; tax credit incentives 221; *see also* Hurricane Katrina

New Orleans Superdome 40, 71
New Public Management (NPM) 133–4
New York City, Hurricane Sandy recovery program 48–9
New York City (Santora & Weiser 2013) lawsuit 73
New York Times 201
Nigro, N.G. and Kellough, J.E. 134
NIMS (National Incident Management System) 152, 157
non-employer business 106
non-essential employees 138, 140, 149, 150, 162
normal accident theory 120
North American Industry Classification System (NAICS) 104, 111; Office of Management and Budget 111
Northridge, California 54
Northrop Grumman Ship Systems 42
Northwest Cocos 176
no-year account 184
Noy, I. 211

Obama, President Barack 39, 47; appointment of czars 52–3
occupational licencing 98
Offering Circular Supplement 178
Office of Foreign Disaster Assistance (OFDA) 186–7; coordination of disaster relief to foreign governments 190–1
Office of Management and Budget 111
Office of State and Local Government Coordination and Preparedness 182
OIC (Officer in Charge) 8, 46, 50–5; peacetime responsibility 37; responsibilities and authority of 37
oil spill: federal government disaster declaration 6; Gulf 4, 6
Orascom Construction Industries 220
organisational behavior 143
organizational strategy 141
Original Principal Amount, cat bonds 177

Parametric Trigger Design 175–6, 177, 190
Pascagoula 40
passive mitigation 116

Patriot Cat Bond Fund (PCBF) 190
Pendleton Act (1883) 133
people with disabilities: blind children 61; deaf children 61; discrimination against 62; historic struggles of 61–4; legislation affecting 62–4; location emergency action plans 76; mentally ill 61–2; out of sight, out of mind approach 62; proactive planning 75; protection from discrimination 67; self-sufficiency planning 76; soldiers 62; *see also* AFN (access and functional needs)
personnel 132; reform as a result of crisis 135–8; *see also* employees; human capital
Pfeiffer, J. and Salancik, G.R. 196
Picou, J.S. and Martin, G.C. 211
Pike *et al.* 197
policy *see* public policy
political patronage 133
politicians, local 157–8
population growth 169, 200, 204
post-disasters: citizen inaction 88; collective-action problem 88; difficulties facing government post-Katrina 91; disaster assistance from top down 90–3; enabling local communities 98; federal government's optimism about capabilities 97–8; freedom of communities to begin rebuilding process 97–8; government inaction 88; government rules and regulations 98; initiatives of local entrepreneurs and community leaders 88, 89, 94–6, 98; the knowledge problem 90; realistic and achievable outcomes 97–8; recovery from the bottom up 93–6; transfer of pre-disaster authority and decision-making to local authorities 97–8; *see also* disaster management; emergency management
Post-Katrina Emergency Management Reform Act (PKEMRA, 2006) 38, 73, 90, 91–2; employees salaries 131
Powell, C. 50
Powell, D. 55n3
preparedness: checklist elements 113; collaboration 154, 155; creating an environment of 158–9; emergency management phase 38, 65, 66, 75, 79, 89, 92, 113–15; FEMA (Federal Emergency Management Agency) 92; public policy 92, 93; *see also* emergency-preparedness
Presidential Policy Directive on National Preparedness (PPD-8) (2011) 92
President's Commission on Employment of the Handicapped (1958) 63
President's Czars, The (Sollenberger and Rozell) 53
President's Management Agenda (PMA) (2001) 135
procurement *see* public procurement
production, cost of 200
property market: concentration in coastal areas 200; development of 200; financial impact of disasters 200; investment in infrastructure 204, 205; population density 200; stricter building standards 204; taxes 204–5
property tax 131, 214, 215
Public Assistance (PA) funds 43–4
Public Assistance Technical Assistance Contracts 29
Public Financial Management 48
public interest: awarding of contracts to established companies 30; contested space 19; definition 19; and public procurement 16
Public Law 107-296 (107th Congress) 136
public-management theory 196
public policy: anchoring expectations 98; centralization 92; definition of small business 103, 112; implications for, small businesses 120–1; nation-wide preparedness goals and metrics 93; preparedness, focus on 92; reform of 154
public procurement: competitive practices 15; contracts awarded to established businesses 29–30; favouring established businesses 29, 30; full and open competition 16–19, 24–5, 26–8, 31; improving 18; sole source 18; urgency 18–19, 26, 29, 30

public sector, impact of Hurricane Sandy on employment 24
public servants: fitness of character 132; political patronage 133; spoils system 132–3
Putnam, R. 36

Raddatz, C. 226
Rapid Repairs program 48
recovery: contrast with resiliency 196; economic growth and 210–11; emergency management phase 29, 36, 51, 55, 112–13, 137–8; Hurricane Katrina 41, 45, 46, 47, 48, 49; Hurricane Sandy 48–9, 50; local *see* local recovery; OIC proposals 51–2; *see also* disaster recovery
registries, special needs 72–4; communications 77; current issues with 74–6; education 78; independence 77; medical care 78; supervision 77; transportation needs 77
Rehabilitation Act (1973) 63
reinsurance 174–5; catastrophe 187
relief spending 4–5
Renne, J. 71
research: categorizing business by activity type 112; definition of small business 112; emergency management, business mitigation activities 115–16; emergency management, business responses 116–17; emergency management of businesses 117–19; emergency management, preparedness of businesses 113–15; emergency management, recovery of businesses from disaster 112–13; implications for, small businesses 120; on small business 112–19
resiliency: concept of 196–7; contrast with recovery 196; definition 195, 196; economic definition 197; psychological perspective 196; *see also* financial resiliency
response, emergency management phase 29, 30, 35–7, 51, 116–17, 137–8
Revenue bonds 203
revenue forecasts 204

Riley Act (1933) 139
Road Home initiative 45–6
Roberts, P.S. 4
Robert T. Stafford Disaster Relief and Emergency Assistance Act *see* Stafford Act (1988)
Rogoff, P. 46
Roosevelt, President T. 181
Ryan, R.W. 135

Sadiq, A. and Weible, V. 116
sales tax 199, 209, 213, 214–15; City of Beaumont 216, 218–20, 221, 222, 225–6; City of Waco 216, 218–20
Sandy Recovery Improvement Act (2013) 39
San Francisco earthquake (1906) 181
Scavo *et al.* 3
schools: for blind children 61; for deaf children 61
Schwab, A.A. and Beatley, T. 66
seismic risks 201
Selden, S.C. 134
September 11 attacks 1–2, 3, 6, 7, 135
service animals 73
Shiramizu, S. and Singh, A. 134
Simplified Acquisition Procedure (SAP) 26
Skidmore, M. and Toya, H. 223
skills inventories 150
Small Business Act (1953) 103
Small Business Administration (SBA) 103–4; defining HUBZones 108; defining small businesses 104; Office of Advocacy 104
small businesses: awarding contracts to 20; bidding for post-disaster contracts 15; categorizing by activity 111; categorizing by emergency management research 111; community risk profile 121; contracts post-Hurricane Sandy 26; as defined by federal government 31, 104; as defined by the SBA 104; as defined in Maryland 104–6, 108; as defined in Montgomery County, Maryland 105–6; defining 103–11; definitions in policy and research 112; disaster literature 102; economically and

socially disadvantaged 106–7, 108; emergency management research 108; future research on 120; gender and 108; geography and 108; implications for emergency management practice on 121–2; implications for policy on 120–1; as key component of the U.S. economy 102; mitigation activities, emergency management research on 115–16; mitigation measures 119; non-employer businesses 106; preparedness, emergency management research on 113–15; preparedness measures 119; recovery after disaster, unlikelihood 119; recovery from disaster, emergency management research on 112–13; reopening after disasters 211–12; responses, emergency management research on 116–17; as a vulnerable population 118–19
Small Disadvantaged Business program 106
Smith, C.H. 21
Sobel, R. and Leeson, P. 90
social capital 36, 37; bonding 93, 94; bridging 93, 94; correlations with 93; linking 93–4
socially disadvantaged individuals, legal definition 107
Social Security Act (1935) 62
social vulnerability 201; theory 118
Soil Conservation Districts (SCDs) 139, 161
sole sourcing 18
Sollenberger, M.A. and Rozell, M.J. 53
South East Texas Disaster Relief Fund 221
South East Texas Emergency Relief Fund (SETERF) 222
South Texas Economic Development Foundation (STEDF) 221, 222
special needs: daily challenges/needs 70; definition 64; evacuation planning 71–2; issues during Hurricane Katrina 67–72; late evacuation during Hurricane Katrina 68; lawsuits arising from Hurricane Katrina 73; location emergency action plans 76; registries 72–4; registries, current issues with 74–6; transportation needs 68, 69; *see also* AFN (access and functional needs); people with disabilities; registries, special needs
Special Purpose Vehicle (SPV) 177
spoils system 132–3
spontaneous, unaffiliated volunteers (SUV's) 128
Stafford Act (1988) 3; business needs and government responses 19–21; contracting expenditures 15; FEMA PA funds 43; good intentions succumbing to pragmatism 29; powers of FEMA 6; purpose of 38; review of 38–9; titles of 38; *see also* public procurement
Standard Industrial Classification (SIC) code system 111
Stark, J. 204
state governments: evaluating fiscal condition of local governments 199; experience with human capital management and disasters 138–9; financial responsibility for disaster response and recovery 198; human resource changes 134; increasing financial resiliency 198
states (U.S.), leadership 36
St. Bernard Parish 95
storms *see* individually named hurricanes
storm surge 21
Storr, V.H. and Haeffele-Balch, S. 91
strategic agility 141, 148
strategic direction 148–51
Strategic Human Capital Management (SHCM) 136
strategic sensitivity 141
Strömberg, D. 223
Sullivan, J. 3
Sundquist, J.L. and Davis, D.W. 3
Superfund Amendments and Reauthorization Act of 1986 (SARA Title III) 146
support agencies 75
support personnel 149
Sutter, D.S. i
Swiss Re Capital Markets 176, 177
Sympathetic State (Dauber) 5

tabletop exercises 147

Tadelis, S. 29
Tenth Amendment 3
Texas: frequency of natural disasters 213–14; tax revenues 214
Texas Comptroller of Public Accounts 214, 221
Texas Disaster Relief Fund 221
Texas Engineering Extension Service (2010) 215, 220
Texas Windstorm Insurance Association (TWIA) 202
Thompson, F. and Miller, H.T. 133–4
Tierney, K.J. 102, 109, 113, 120
Times-Picayune, The 68, 70
top down assistance 90–3
tornadoes 87; Calhoun County, Alabama 127; financial costs of 170
tourism industry, response actions within 117, 118
Toya, H. and Skidmore, M. 201, 209, 213
training 149–50, 152, 155–6, 157
Treasury, U.S.: issuing of cat bonds 189–90; obtaining funding for disaster relief 188–9; Office of Emergency Programs 189; Patriot Cat Bond Fund (PCBF) 190
Tropical Storm Fay 71

United States Civil Service Commission 133
unity of effort 141
University of Illinois 63
Urban Mass Transit Act (1970) 63
Urban Mass Transportation Act (1964) 63
urgency, public procurement 18–19, 29, 30; subject to interpretation 30
URS Corporation 48
USAID (United States Agency for International Development): obtaining funding for disaster relief 188; Office of Foreign Disaster Assistance (OFDA) 186–7, 190–1
U.S. Census Bureau Tax Foundation (2013) 214
U.S. Department of Housing And Urban Development (HUD) 41, 45
U.S. Department of Housing and Urban Development, report on Hurricane Sandy 21
U.S. Geological Survey 54, 176
U.S. National Hurricane Center 176
U.S. Securities and Exchange Commission 111

Vien, Father 94, 94–5
Vocational Rehabilitation Amendment (1943) 63
Voitier, D. 95–6
Voluntary Organizations Active in Disasters (VOAD) 146–7, 161
volunteer-registration centers 128
volunteers 137–8; as disaster service workers in California 138–9; management of 150–1, 151; *see also* spontaneous, unaffiliated volunteers (SUV's)
vulnerable populations 73; education for 78; small businesses as 118–19

Waco case study 215–16; economic characteristics 216; fertilizer explosion 215; housing and employment characteristics 216; sales tax revenue 216, 218–20
War Powers Act (1943) 138
Washington, DC (Beck 2014) lawsuit 73
wealth, financial vulnerability and 201
Webb *et al.* 113, 122
Whole Community concept 2, 89, 92
Women-Owned Small Businesses (WOSB) 108
Women-Owned Small Business Federal Contract Program 106
Woolcock, M. 93
World Bank: Department of Treasury 176, 177; disaster relief to foreign governments 190–1; MultiCat program *see* MultiCat Mexico 2009 Catastrophe Bond; MultiCat program

Yoshida, K. and Deyle, R.E. 110, 116, 121